An Introduction to
Geological Structures
and Maps

Eighth Edition

An Introduction to Geological Structures and Maps

Eighth Edition

Dr George M. Bennison
Chartered Geologist, formerly Senior Lecturer in Geology,
University of Birmingham, UK

Dr Paul A. Olver
Chartered Geologist, formerly Lifelong Learning
Development Officer, Herefordshire Council, UK

Dr Keith A. Moseley
Head of Physics and Geology, Monmouth School, UK

Routledge
Taylor & Francis Group
New York London

First edition published in Great Britain in 1964
Eighth edition published in Great Britain in 2011 by Hodder Education

First published 2013 by Routledge

2 Park Square, Milton Park, Abingdon, Oxon OX14 4RN
711 Third Avenue, New York, NY, 10017, USA

Routledge is an imprint of the Taylor & Francis Group, an informa business

British Library Cataloguing in Publication Data
A catalogue record for this title is available from the British Library

Library of Congress Cataloging-in-Publication Data
A catalog record of this book is available from the Library of Congress

ISBN 13: 9781032320182 (pbk)

Typeset in 10 on 12 pt Palatino by Phoenix Photosetting, Chatham, Kent
Cover image: The Green Bridge, Pembrokeshire, Wales © Paul A. Olver

Printed in the UK by Severn, Gloucester on responsibly sourced paper

Contents

List of Plates

Preface to the current edition

As with previous editions of this book, the 8th edition seeks to cover topics up to and including first year undergraduate level. With many earth science students meeting the subject for the first time at university, it also includes the basic principles of geology, first laid down in the 19th Century, which are fundamental to the study of geological structures in the field.

Fieldwork experience is a vital component in all geological training and this edition, with its additional descriptions and exercises on cliff exposures and quarry faces, seeks to relate what is seen in the field in three dimensions with its representation on a geological map,

The inclusion of colour photographs for the first time reinforces this fieldwork connection by showing a variety of geological structures at outcrop which can then be related to the map exercises. Additions to the text, such as way-up criteria, further develop this important fieldwork theme.

Chapter 11 on Igneous and Impact features has been considerably expanded to emphasise the importance of these processes in the study of planetary surfaces. Plate tectonics and sea-floor spreading are highlighted on Map 30 based on an ophiolite complex in Oman while the recognition of large-scale impact structures on our own planet is featured on Map 36.

I would like to thank Dr George Bennison for inviting me to be a co-author on this edition. My first association with the book was with its first edition whilst studying 'A' Level Geology at school, a process that continued with my undergraduate studies at the University of Birmingham with Dr Bennison as my course tutor. I thus feel very much part of its evolution and Dr Bennison's constant support and useful discussions on key topics have ensured that my contribution has been successfully integrated into the whole.

I would also like to thank Dr Keith Moseley for his contribution particularly his excellent photographs now seen at their best in full colour. Further thanks are due to the late Peter Thomson, who first introduced me to the details of Herefordshire's geology, and to Diana Smith, a colleague on many geological excursions in Britain and abroad, for the use of their photographs.

Dr Paul Olver
Canon Pyon
Herefordshire
January 2011

Preface to previous editions

This book is designed primarily for university and college students taking geology as an honours course or as a subsidiary subject. Its aim is to lead the student by easy stages from the simplest ideas on geological structures right through a first year course on geological mapping, and much of its content will also be relevant to students of 'A' Level Geology. The approach is designed to help the student working with little or no supervision; each new topic is simply explained and illustrated by Figures with exercises being set on the associated problem maps. If students are unable to complete the problems they should read on to obtain more specific instructions on how the theory may be used to solve the problem in question. Completed sections for every map (where a section is required) together with the answers to other questions associated with both maps and problems are provided in an Appendix.

In addition to problem maps based on, or adapted from, published geological maps, reference is also made in each Chapter to British Geological Survey maps (generally on the 1:50,000 scale). They are specifically selected to best illustrate the key content of each Chapter. Some of the early maps in the book are of necessity somewhat 'artificial' so that new structures can be introduced one at a time thus retaining clarity and simplicity.

Structure contours (see p.13) are seldom parallel in nature; it is therefore preferable to draw them freehand, though – of course – as straight and parallel as the map permits. In all cases except the 'three-point' problems, the student should examine the maps and attempt to deduce the geological structures from the disposition of the outcrops in relation to the topography, as far as this is possible, before commencing to draw structure contours.

A recent trend noted in syllabuses and degree modules covering structural geology is the reduction in the use of structure contours in understanding maps and solving structural problems. Two Chapters, on *Map Solution without Structure Contours*, have therefore been introduced at different stages in the book.

The authors wish to record their appreciation of all the support given by the late Dr. F. Moseley in the development of this book.

Key to Maps

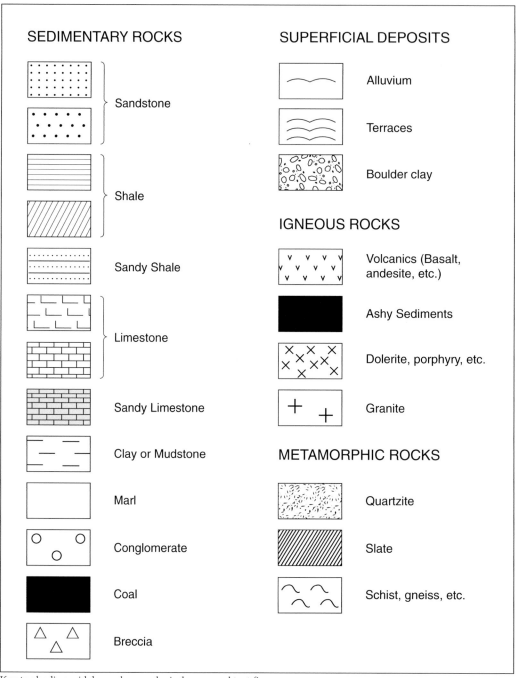

SEDIMENTARY ROCKS

Sandstone

Shale

Sandy Shale

Limestone

Sandy Limestone

Clay or Mudstone

Marl

Conglomerate

Coal

Breccia

SUPERFICIAL DEPOSITS

Alluvium

Terraces

Boulder clay

IGNEOUS ROCKS

Volcanics (Basalt, andesite, etc.)

Ashy Sediments

Dolerite, porphyry, etc.

Granite

METAMORPHIC ROCKS

Quartzite

Slate

Schist, gneiss, etc.

Key to shading widely used on geological maps and text figures.

Strata – the ground rules

The onset and rapid spread of canals across the face of Britain in the late eighteenth century, closely followed by the building of the railway network in the early nineteenth century, were largely responsible for making the study of the strata, or stratigraphy, a subject of both practical and economic value.

It is not surprising, therefore, that it was a land surveyor and canal engineer, William Smith (1769–1839), working initially in southern and eastern England, who first worked out that rock strata were not randomly disposed around the country but arranged in a definite order. He recognised that sedimentary rocks, 'the sediments of past ages', first identified as such by James Hutton (1726–1797), are laid down with the oldest sediments at the base of the sequence and progressively younger beds resting on top. This was defined as the Law of Superposition (Fig. 1.1), and led to Smith being called 'Strata Smith' by his contemporaries.

His work, particularly in the Cotswold Hills on the construction of new canals, led him to observe that each particular stratum of rock yielded its own distinctive assemblage of fossils. In addition, these fossil assemblages always occurred in the same order in different parts of the country. Furthermore, in some parts of the country, the rock type or lithology might change but parts of his recognised fossil assemblages remained and therefore the rock strata could be correlated despite the lithological variations.

So strata can, as Smith put it, be 'identified by their organised fossils' and he soon derived the second concept, the Principle of Strata Identified by Fossils. The characteristics seen by Smith pre-dated Darwin's doctrine of evolution, which was to appear in 1859, and they are now seen as the result of organic evolution whereby particular species change into new and different forms. With the passing of geological time, organisms become extinct and never reappear. This is known as the Law of Biotic Succession (Fig. 1.2).

Therefore, each fossil animal or plant can be used to define a particular geological time interval and, the shorter that span of time is, the more useful the fossil becomes in dividing up geological time. Those fossils that are particularly diagnostic are known as zone fossils. This method of using fossils is defined as biostratigraphy and can be compared to lithostratigraphy, where divisions

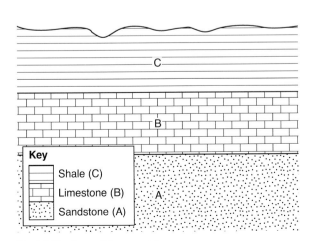

Fig. 1.1 The Law of Superposition.

Key
Shale (C)
Limestone (B)
Sandstone (A)

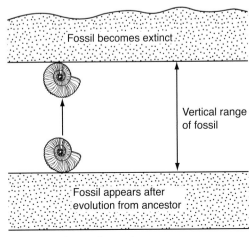

Fossil becomes extinct

Vertical range of fossil

Fossil appears after evolution from ancestor

Fig. 1.2 The Law of Biotic Succession.

within the stratigraphic column are defined by rock types alone. (See Fig. 1.5 for nomenclature.)

The dating of rocks by identification of fossil assemblages is defined as relative dating. Absolute dating of a rock in years cannot be achieved by fossils and, since the late 1930s, has been achieved by radiometric methods. This relies on the presence of radioactive isotopes, such as K^{40} or C^{14}, which occur naturally in small quantities in rocks, to undergo decay at a certain rate to form 'daughter' isotopes, usually of another element. By knowing the decay rate, or half-life, of an individual radioactive isotope, the time since the initiation of decay processes, such as the crystallisation of an igneous rock, can be calculated. Only certain rock types, mainly igneous rocks formed by the crystallisation of molten melts or magmas, can be successfully radiometrically dated. However, the accumulative results have now produced a series of geological periods whose beginnings are given in millions of years (Fig. 1.5). This form of stratigraphy is called chronostratigraphy. Two further important concepts now need to be considered. In many cases, sedimentary rocks, particularly coarse-grained conglomerates and breccias, contain fragments of other rocks. By definition, these rocks must be older than the enclosing sediments, which underlies the basic premise of the Law of Included Fragments (Fig. 1.3).

An important assumption in structural geology is that, in the main, a series of sedimentary rocks is deposited on a surface that is close to being the horizontal. Exceptions to this rule are scree deposits – coarse sediments laid down by flash floods in a desert or storm event erosion of a coral reef. Another example is the building of delta sands into deep water. These localised situations apart, which are usually clearly identifiable, sedimentary rocks are generally deposited with virtually no dip and the Principle of No Initial Dip can be formulated.

This means that the occurrence of large-scale non-horizontal geological strata is the result of earth movements, rather than of sedimentary processes.

Finally, if a rock sequence is cut by an igneous intrusion, such as a dyke, then the intrusion must be the youngest rock (Fig. 1.4). The relationship between geological successions and different types of intrusions will be explored further in Chapter 11.

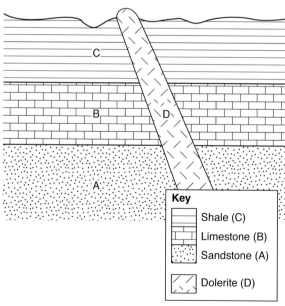

Fig. 1.4 The Law of Cross-Cutting Relationships.

Key

Shale (C)
Limestone (B)
Sandstone (A)
Dolerite (D)

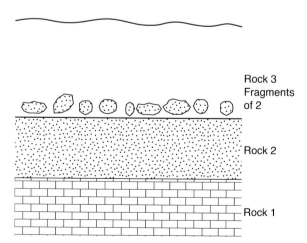

Rock 3
Fragments of 2

Rock 2

Rock 1

Fig. 1.3 The Law of Included Fragments.

Ma	EON	ERA	PERIOD	
			QUATERNARY	
1.6				
5.3		CAENOZOIC		PLIOCENE
				MIOCENE
23			TERTIARY	OLIGOCENE
34				
56				EOCENE
65	PHANEROZOIC EON			PALAEOCENE
145		MESOZOIC	CRETACEOUS	
200			JURASSIC	
248			TRIASSIC	
299		UPPER PALAEOZOIC	PERMIAN	
360			CARBONIFEROUS	
417			DEVONIAN	
443		LOWER PALAEOZOIC	SILURIAN	
490			ORDOVICIAN	
545			CAMBRIAN	
2500	Proterozoic		PRECAMBRIAN	
4000	Archaean			
4600	Hadean			

Fig. 1.5 The subdivisions of geological time.

Horizontal strata

In the simplest case, we can consider a series of strata whose outcrop is horizontal. Rarely are they so in nature; they are frequently found elevated hundreds of metres above their position of deposition, and tilting and warping have usually accompanied such uplift. The pattern of outcrops of the beds where the strata are horizontal is a function of the topography; the highest beds in the sequence (the youngest) will outcrop on the highest ground and the lowest beds in the sequence (the oldest) will outcrop in the deepest valleys. Such a horizontal orientation is rare in Britain but would be typical of road cuttings, say, in the Great Plains of western USA (Fig. 2.1). The individual beds are separated by bedding planes. These are successive surfaces of deposition and represent the earth's surface during the time between individual sedimentary deposition phases. The outcrop contains no other geological features except for a few joints, formed either through tension during its vertical uplift or by decompaction as former sedimentary layers above have been removed.

The hills in valleys of the landscape, in which layered sequences of sedimentary rock outcrop, display their differing thicknesses and resistance to

Plate 1 Horizontal bedding. View from the South Rim, Grand Canyon, Arizona. This large erosional feature exposes horizontally bedded sedimentary rocks (also see Map 1).

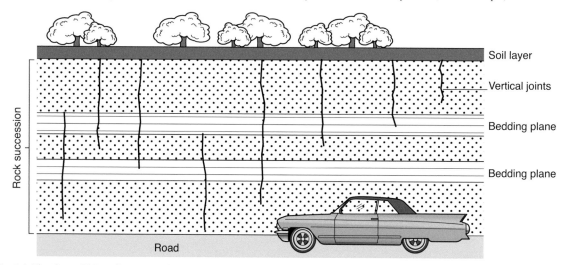

Fig. 2.1 The Great Plains of western USA – a typical roadside section.

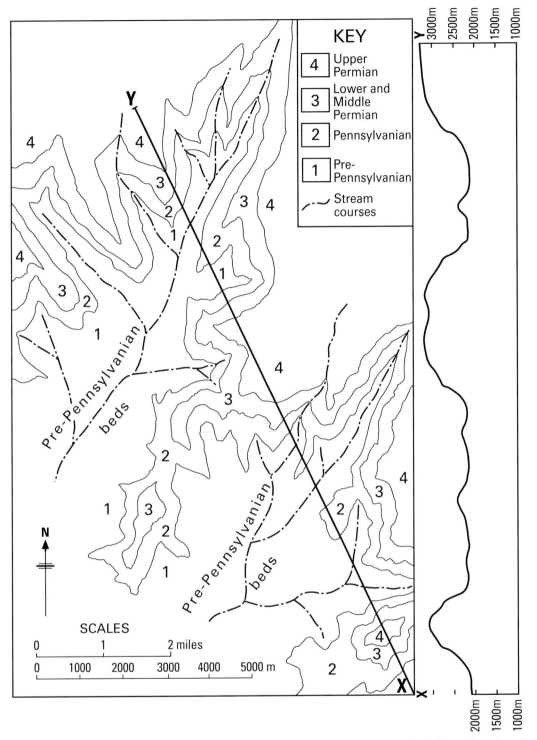

KEY

4 Upper Permian

3 Lower and Middle Permian

2 Pennsylvanian

1 Pre-Pennsylvanian

Stream courses

SCALES

0   1   2 miles

0   1000   2000   3000   4000   5000 m

Map 1 This shows a simple geological map based on the geology of the Bright Angel area of the Grand Canyon, Arizona. The strata are horizontal (see Plate 1). How can horizontal strata give rise to such a complicated pattern of outcrops? It is because the topography is complicated. A plateau has been very deeply dissected by a dendritic pattern of rivers and tributaries. The high altitude of this region and low base level give the streams great erosive power, which in this fairly arid region of sparse vegetation has resulted in steep-sided canyons with many branches. You may find it helpful to colour the different strata to make the outcrop pattern clearer. For simplicity, contours have been omitted from the map. Of course, the outcrops of the geological boundaries between strata of different ages are themselves contours, since the strata are horizontal.

The base of the Pennsylvanian is at a height of 2250 m and it is 500 m thick. The Lower and Middle Permian together are 250 m thick. Insert the geology on the profile provided along the line of section X–Y.

Plate 2 Jointing in siltstones (parallel to hammer handle) in upper Jurassic Kimmeridge Clay. Near Freshwater Steps, Dorset.

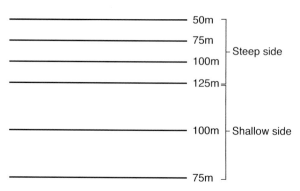

Fig. 2.3 A typical ridge.

erosion by their associated land forms and diverse topography. Only in exceptional circumstances is the topography eroded out of a single rock type. Topographical maps produced by the Ordnance Survey and others show the form of the ground by means of contours.

Contours

A contour is a line that is, at every point, the same height above a reference level, sea level (defined at Newlyn in Cornwall), and horizontal. A simple uniform slope would have contours at equally spaced intervals (Fig. 2.2). The steeper the slope, the closer is the spacing of the contours. Consider a ridge, or cuesta, with a steep side and a less steeply dipping side (Fig. 2.3). The contours on the steep

side will be closer together than those on the less steeply sloping side. A hill, standing alone above the general level (such as Bredon Hill, Worcestershire), will have concentric contours, seldom perfectly circular in nature, of course (Fig. 2.4). A typical valley produced by erosion by a stream or

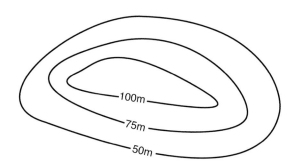

Fig. 2.4 An isolated hill.

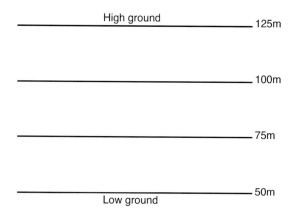

Fig. 2.2 A uniform slope.

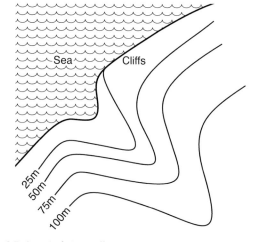

Fig. 2.5 A typical river valley.

river will be V-shaped in section and its contours will be V-shaped as in Fig. 2.5, which shows the contours of a stream near Zennor in Cornwall. Looking at a contour map we can envisage the complete topography of the area of the map.

Horizontal and vertical strata in the landscape

Fig. 2.6 shows a series of horizontal strata of varying resistance to erosion outcropping within a river valley. The heights above sea level are shown in the form of contours superimposed on the landscape. It can be seen that where geological strata are horizontal, the relationship to the topography is very simple. The pattern of the geological boundaries is dependent on the contours and the geological boundaries or contacts (since they are horizontal) will be parallel to the ground contours and will themselves be contour lines.

Fig. 2.7(a) shows a similar situation on a cliff coast where a horizontal sequence of conglomerate, limestone and sandstone has been intruded at a later date by a dolerite dyke of igneous origin. This dyke is vertical and cuts through the sedimentary contacts at right angles.

Looking down from above (in the direction of the arrows), the geological map of this area would be revealed in Fig. 2.7(b). Despite the top of the cliffs being uneven in terms of topography, the igneous dyke outcrops as a straight line across the map and its contours.

Geological maps

A geological map generally shows the geological strata that are present below the vegetation and soil (and superficial deposits). These are drawn on to a topographical map with contours that show the form of the ground. The geologist producing a map observes all the geological features that can be seen, in cliffs and escarpments and in minor outcrops, in man-made exposures such as quarries and road cuttings, and valleys where the river or stream has cut down to reveal geological features – as in a ravine. He locates his position and the

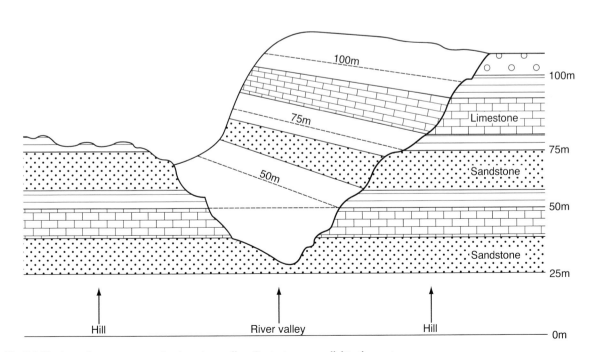

Fig. 2.6 Horizontal strata outcropping in a river valley. Contacts are parallel to the contours.

position of all these geological features on a topographical map, or sometimes on an aerial photograph. He can now deduce the complete geology of the area by interpreting and collating all the information. Problem Maps 44 and 45 are examples of 'field maps' produced in this way. When you have worked through the exercises up to that point in the book you will be able to complete the geology of those areas. Map 35 is a simplified version of the BGS (British Geological Survey) map of the Isle of Arran. Of course, in many areas the surface observations of the geologist are supplemented by information from boreholes. Several map problems here are based on borehole information. We are fortunate in Great Britain to have examples of geology from most of the geological record, that

is to say, from all geological periods. Also we have examples of all types of sedimentary, igneous and metamorphic rocks. In our long and complex geological history, mountain-building episodes have resulted in a very wide range of geological structures. In some parts of the world the strata have been uplifted from the site of their deposition with very little disturbance so that the strata are horizontal, or very nearly so, for considerable distances, as in the Great Plains of the USA (Fig. 2.1). A geological map would be very simple. An example of nearly horizontal strata that have been very deeply eroded is where the Colorado River has eroded the mile-deep Grand Canyon and so the great succession of strata is exposed. Map 1 is based on a small area of the Grand Canyon called Bright Angel.

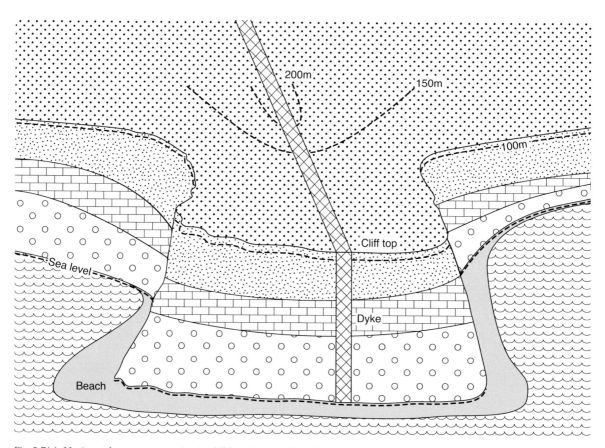

Fig. 2.7(a) Horizontal strata outcropping in cliff face showing the relationship between the landscape and the resultant map. Note relationship of dyke with sedimentary strata.

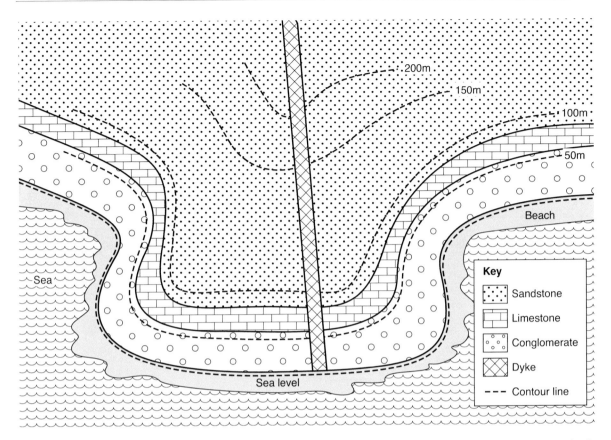

Fig. 2.7(b) Geological map of the same area as Fig. 2.7(a). Note that the vertical dyke produces a straight outcrop across a landscape of varying heights above sea level.

The maps in this book are based on real geological structures but may be simplified to illustrate a particular point. Starting with the simplest situation – where beds are horizontal – more complicated structures are introduced and the maps are more realistic; the geological area shown in Map 45 exists in the north-west of England in the Cross Fell inlier and an area of Oman is represented on Map 30.

Section drawing

Draw a base line the exact length of the line A–B on Map 2 (19.0 cm). Mark off on the baseline the points at which the contour lines cross the line section: for example, 8.5 mm from A, mark a point corresponding to the intersection of the 700 m contour. From the baseline, erect a perpendicu-

lar corresponding in length to the height of the ground and, since it is important to make vertical and horizontal scales equal wherever practicable, a perpendicular of length 14 mm must be erected to correspond to the 700 m contour (since 1000 m = 2 cm and 100 m = 2 mm) (Fig. 2.8). Sections can readily be drawn on metric squared paper (or on 1/10″ in some cases).

Map 3 shows an area of the Cotswold Hills and adjacent lowlands, where the strata are virtually horizontal. Geological boundaries, therefore, are parallel to topographic contours, a point made on p. 7. Contours have been omitted for clarity but the general heights of the hills and plain are given. These, together with the altitude of a number of points (spot-heights and triangulation points), enable us to draw a sufficiently accurate topographic profile of a section across the map.

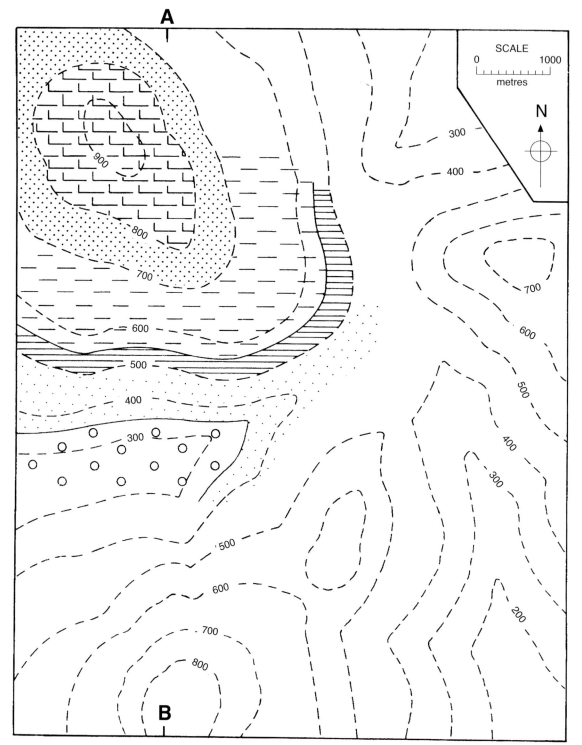

Map 2 The geological outcrops are shown in the north-west corner of the map. It can be seen that the beds are horizontal as the geological boundaries coincide with, or are parallel to, the ground contour lines. Complete the geological outcrops over the whole map. Indicate the position of a spring-line on the map. How thick is each bed? Draw a vertical column showing each bed to scale: 1 cm = 100 m. Complete a section along the line A–B. (Contours in metres.) See Fig. 2.8.

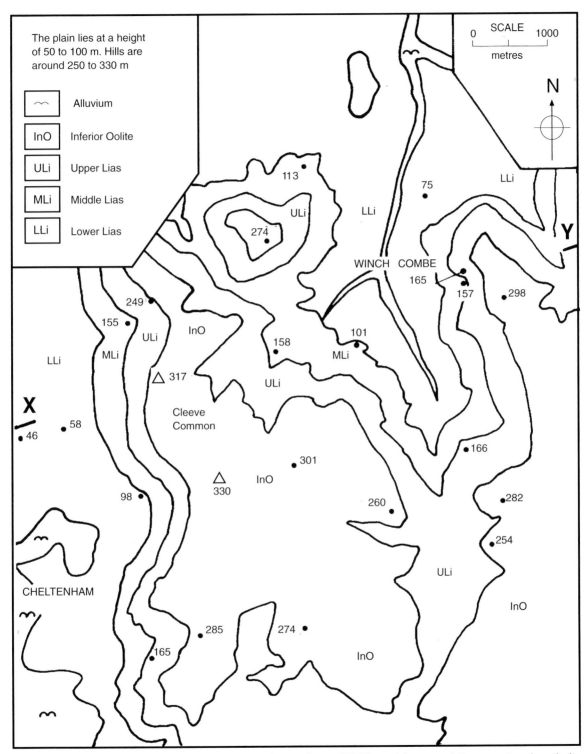

Legend:

The plain lies at a height of 50 to 100 m. Hills are around 250 to 330 m

- Alluvium
- InO — Inferior Oolite
- ULi — Upper Lias
- MLi — Middle Lias
- LLi — Lower Lias

SCALE 0 — 1000 metres

N

Y

113

75

LLi

ULi

LLi

WINCH COMBE

274

165

157   298

249

155   ULi   InO

158   101

MLi

LLi   MLi

△ 317   ULi

X   Cleeve Common

58

46

△ 330   InO

301

98

260   282

254

166

ULi

CHELTENHAM   InO

285   274

165   InO

Map 3 This map is based on a portion of the British Geological Survey map of Moreton-in-the-Marsh, 1:50,000 scale, Sheet 217. (It is slightly simplified and reduced to fit the page size.) Reproduced by permission of the Director, British Geological Survey: NERC copyright reserved. Draw a section along the line X–Y to illustrate the geology, using a vertical scale of 1 cm = 200 m.

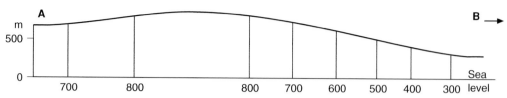

Fig. 2.8 Part of a section along the line A–B on Map 2 to show the method of drawing the ground surface (or profile).

Vertical exaggeration

The horizontal scale of a map has been determined at the time of the original survey. Commonly, we find it given as 1:50,000 – 1 cm on the map representing 50,000 cm or 500 m on level ground (1:50,000 is the 'Representative Fraction'). Maps on the scale of 1:25,000 and 1:10,000 are common. On older maps one inch represents one mile (1:63,360) and on USA maps two miles to the inch is a usual scale.

If a similar scale to the horizontal is used vertically, we often find that the section is difficult to draw and that it is very difficult to include the geological details. For example, on Map 2 the highest hill on the section at 317 m would be only just over 0.6 cm above the sea-level baseline. A suggested vertical scale of 1 cm = 200 m, with horizontal scale predetermined as 1 cm = 500 m, gives a vertical exaggeration of 500/200 or 5/2 or 2.5.

Complications arise where the strata are inclined and further considerations of vertical exaggeration are discussed on p. 19.

When you have drawn the geological section on the profile provided, turn to p. 140 in the Appendix, where the geological section has been drawn on a true scale (horizontal and vertical scales the same) in Fig. A1. It has also been redrawn (Fig. A2) with a vertical exaggeration of × 5 (horizontal scale 1 cm + 500 m, vertical scale 1 cm = 100 m). The true scale section in Fig. A1 illustrates the practical difficulty, while Fig. A2 shows that too great a vertical exaggeration produces an unnatural distortion (and in the case of dipping strata causes other problems, dealt with at the end of this chapter). Where vertical exaggeration is necessary from a practical point of view, it should be kept to a reasonable minimum.

This topic is discussed again in Chapter 7, with reference to Map 15.

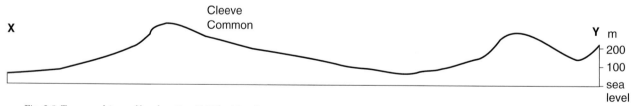

Fig. 2.9 Topographic profile of section X–Y for Map 3.

3

Dipping strata

Inclined strata are said to be 'dipping'. The angle of dip is the maximum angle measured between the strata and the horizontal (regardless of the slope of the ground) (Fig. 3.1 and Plate 3).

The direction of dip is given as a compass bearing reading from 0° to 360°. For example, a typical reading would be 12/270. The first figure is the angle of dip, the angle that strata make with the horizontal. The second figure is the direction of that dip measured round from north in a clockwise direction (in this example due west). In North America, where the use of the Brunton Compass is almost universal, dip directions are more commonly given in relation to the main points of the compass, e.g. 10° west of south (i.e. 190°).

In a direction at right angles to the dip the strata are horizontal. This direction is called the strike (Fig. 3.2 and Plate 4). An analogy may be made with the lid of a desk. A marble would roll down the desk lid in the direction of maximum dip. The edge of the desk lid, which is the same height above the floor along the whole of its length, i.e. it is horizontal, is the direction of strike.

Structure contours (strike lines)

Any surface can be defined by its contours. We are all familiar with 'contours' on maps, by which we generally mean contours (lines joining all points of equal height above mean sea level) that define the topography or ground surface. These lines can be constructed by joining up many points (in theory an infinite number), the height of which has been obtained by conventional surveying methods or from very detailed aerial photographic survey. The latter very costly option may be justified in the construction of major engineering projects such as dams.

Just as it is possible to define topography of the ground by means of contour lines, so any geological surface (but chiefly it is applied to bedding planes) can be defined by a set of contour lines, referred to as structure contours since they reveal and define geological structures. Although your syllabus may not include the constructions of structure contours, the next few pages contain information vital to the understanding of geological structures. These pages include simple mathematics.

Plate 3 A natural arch produced by marine erosion of a dipping Carboniferous Limestone succession. The Green Bridge of Wales, Flimston, Pembrokeshire.

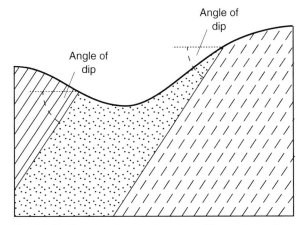

Fig. 3.1 Section showing dipping strata. The angle of dip is measured from the horizontal.

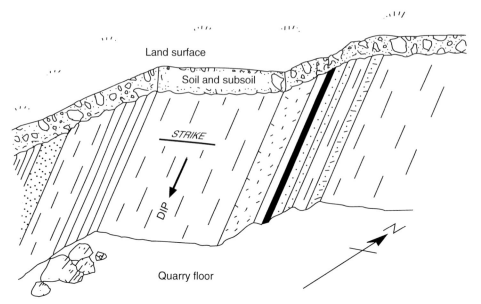

Fig. 3.2 Southerly dipping strata in a quarry. Note the relationship between the directions of dip and strike.

Plate 4 Dip and strike, Silurian strata, Marloes, Pembrokeshire. The cliff shows strata steeply dipping to the right and the wave-cut platform in the foreground exposes the strike of the beds.

Construction of structure contours

Earlier in the chapter the term 'strike' was introduced. This is the direction on a dipping or sloping plane in which there is no component of dip (because it is at right angles (perpendicular) to the True Dip, i.e. the direction of maximum dip). A line drawn in this direction, called a strike-line, is a structure contour since it is drawn on the geological surface and it is horizontal (no component of dip). The terms 'structure contour' and 'strike-line' are synonymous, but for the purpose of this book the term 'structure contour' will be used. Where complex structures occur, the information needed to draw structure contours will be obtained from a programme of boreholes (which will provide a great deal of information as well as the depth to a certain stratum).

As we shall see in Chapter 4, it is usual to convert depths from the surface to heights relative to a horizontal datum, mean sea level commonly being used. This is called Ordnance Datum (O.D.).

In the problems dealing with deep geological structures, such as those encountered in strata beneath the North Sea, some artificial datum (e.g. −1,000 metres below O.D.) may be used.

There is evidence of the height of a geological surface (geological boundary) wherever it can be seen to cross a topographic contour line. For example, the boundary between beds S and T on Map 4 cuts the 700 m contour at three points. These points lie on the 700 m structure contour, which can be drawn through them. Since these early maps portray simply inclined plane surfaces, structure contours will be straight, parallel and – if dips are constant – equally spaced. Having found the direction of strike to be 85°, we know that the direction of dip will be at right angles to this but we must ascertain whether the dip will be 'northerly' or 'southerly'. A second structure contour can be drawn on the same geological boundary S–T through the two points where it cuts the 600 m contour. From the spacing of the two structure contours we can calculate the dip or gradient of the beds (Fig. 3.4(a)).

$$\text{Gradient} = \quad 700 \text{ m}{-}600 \text{ m in } 1.25 \text{ cm}$$
$$\text{i.e.} \qquad\qquad 100 \text{ m in } 1.25 \text{ cm}$$

As the scale of the maps is given as 2.5 cm = 500 m, 100 m in 1.25 cm is equivalent to 100 m in 250 m. Hence, the gradient is 1 in 2.5, to 175°. Frequently, it is more convenient to utilise gradients expressed in this way, 1 in 2.5, 1 in 5, etc., although dips are always given as an angle (to the horizontal) on published geological maps. In a right-angled triangle (Fig. 3.3) the angle of dip is α. The gradient is the vertical difference, A–C, divided by the horizontal distance between A and C measured on the map, i.e. B–C.

$\dfrac{\text{AC}}{\text{BC}}$ = the tangent of the angle α. Substituting values, we have $\dfrac{1}{25}= 0.4$.

Consulting a table of tangents of angles you will see that 0.4 is the tangent of an angle of 22°. In general usage such as steep road gradients, the old-fashioned signs giving values, for example 1 in 10, have been replaced with percentages so that 1 in 10 is given as 10% (1 in 5 = 20%, 1 in

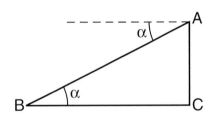

Fig. 3.3 Right-angled triangle showing the relationship to the angle of dip α.

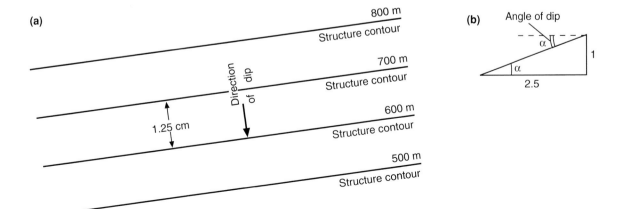

Fig. 3.4(a) Plan showing structure contours and (b) section through contours showing the relationship between dip and gradient.

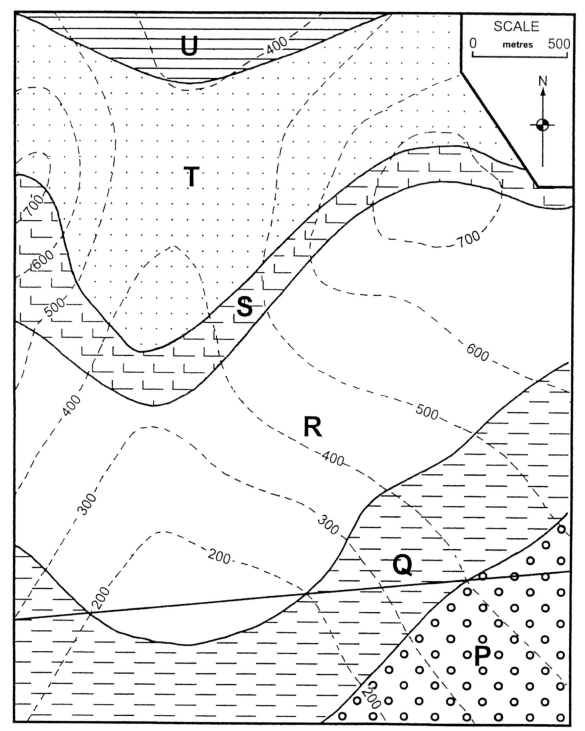

Map 4 The continuous lines are the geological boundaries separating the outcrops of the dipping strata, beds P, Q, R, S, T and U. Examine the map and note that the geological boundaries are not parallel to the contour lines but, in fact, intersect them. This shows that the beds are dipping. Before constructing structure contours, can we deduce the direction of dip of the beds from the fact that their outcrops 'V' down the valley? Can we deduce the direction of dip if we are informed that Bed U is the oldest and Bed P is the youngest bed of the sequence? Draw structure contours for each geological interface[1] and calculate the direction and amount of dip. (Contours in metres.) Instructions for drawing structure contours are given below and one structure contour on the Q/R geological boundary has been inserted on the map as an example.

[1] Some confusion may arise since the term geological boundary is often applied both to the interface (or surface) between two beds and to the outcrop of that interface. It seems a satisfactory term to employ, however, since the two are related and the context generally avoids ambiguity.

8 = 12.5%). To convert a percentage, divide by 100, of course, then divide 1 by this figure.

E.g. $12.5\% = \dfrac{12.5}{100} = 0.125$.

$\dfrac{01}{0.125} = 8$, so gradient is 1 in 8.

Section drawing

The topographic profile is drawn by the method already described on p. 12. The geological boundaries (interfaces) can be inserted in an analogous way by marking the points at which the line of section is cut by structure contours. Perpendiculars are then drawn from the baseline, of length corresponding to the height of the structure contours (Fig. 3.5).

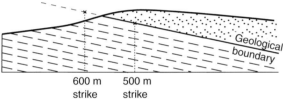

Fig. 3.5 Section to show the method of accurately inserting geological boundaries.

True and apparent dip

If the slope of a desk lid, or of a geological boundary (interface) or bedding plane, is measured in any direction between the strike direction and the direction of maximum dip, the angle of dip in that direction is known as an apparent dip (Fig. 3.6(a)). Its value will lie between 0° and the value of the maximum or true dip. Naturally occurring or man-made sections through geological strata (cliffs, quarry faces, road and rail cuttings) are unlikely to be parallel to the direction of true dip of the strata. What may be observed in these sections, therefore, is the dip of the strata in the direction of the section, i.e. an apparent dip (somewhat less than the true dip in angle). The trigonometrical relationship is not simple:

Tangent apparent dip   = tangent true dip × cosine β

(see Fig. 3.6(b) and 51 and Table 2, Appendix, p. 157).

However, the problem of apparent dip calculation is much simplified by considering it as a gradient. Just as the gradient of the bed in the direction of maximum dip is given by the spacing of the structure contours (1.9 cm = 380 m in Fig. 3.6(b), representing a gradient of 1 in 3.8, since the scale of the map is 1 cm = 200 m), so the gradient in the direction in which we wish to obtain the apparent dip is given by the structure contour spacing measured in that direction (3.25 cm = 650 m in Fig. 3.6(b), representing a gradient of 1 in 6.5).

In road and rail cuttings the direction of dip of strata is vital to the stability of the slopes. Where practicable, a cutting would be parallel to the direction of dip of the strata, minimising

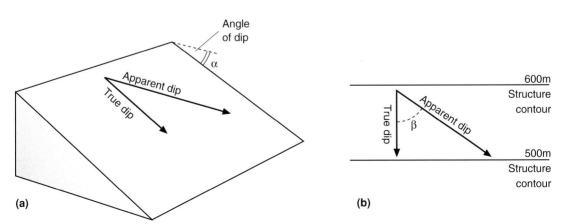

(a)   **(b)**

Fig. 3.6 (a) Diagram and (b) plan or map of structure contours to illustrate the relationship between true and apparent dip.

slippage into the cutting since there would be no component of dip at right angles to the face of the cutting. Factors other than geological ones determine the siting and direction of cuttings. Frequently they are not parallel to the dip of the strata and, as a result, we see an apparent dip in the cutting sides.

In the event of a geological section being at right angles to the direction of dip it will be, of course, in the direction of strike of the beds. There will be no component of dip seen in this section and the beds will appear to be horizontal. (Of course, close examination of such a quarry face or cutting will reveal that the strata are not horizontal but are dipping towards or away from the observer.)

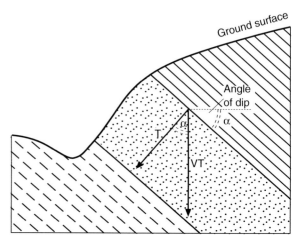

Fig. 3.7 Section showing the relationship between the vertical thickness (VT) and the true thickness (T) of a dipping bed.

Calculation of the thickness of a bed

On Map 5 it can be seen that the 1100 m structure contour for the geological boundary D–E coincides with the 1000 m structure contour for boundary C–D. Thus, along this strike direction, the top of bed D is 100 m higher than its base. It has a vertical thickness of 100 m. This is the thickness of the bed that would be penetrated by a borehole drilled at point X.

Turn back to Map 4. The 200 m structure contour for the Q–R boundary, which has already been inserted on the map, passes through the point where the P–Q boundary is at 400 m. Since structure contours are all parallel – where beds are simply dipping as in this map – this line is also the 400 m structure contour for the P–Q boundary. It follows that bed Q has a vertical thickness of 200 m. You will find that on many problem maps bed thickness is often 100 m, 200 m or some multiple of 50 m in order to produce a simpler problem.

Vertical thickness and true thickness

Since the beds are inclined, the vertical thickness penetrated by a borehole is greater than the true thickness measured perpendicular to the geological boundaries (interfaces) (Fig. 3.7). The angle α between VT (vertical thickness) and T (true thickness) is equal to the angle of dip.

Now cosine $\alpha = \dfrac{T}{VT}$

$\therefore T = VT \times cosine\ \alpha$

The true thickness of a bed is equal to the vertical thickness multiplied by the cosine of the angle of dip. Where the dip is low (less than 5°) the cosine is high (over 0.99) and true and vertical thicknesses are approximately the same (see Table 2, p. 157).

Width of outcrop

If the ground surface is level, the width of outcrop of a bed of constant thickness is a measure of the dip (Fig. 3.8(a)).

Naturally, where beds with the same dip crop out on ground of identical slope, the width of outcrop is related directly to the thickness of the beds (Fig. 3.8(c) and Table 3, Appendix).

More generally, beds crop out (or outcrop) on sloping ground and width of outcrop is a function of dip and slope of the ground as well as bed thickness. In Fig. 3.8(b) beds Y and Z have the same thickness (and dip) but, due to the different angles of intersection with the slope of the ground, their widths of outcrop (W) are very different. ($T_1 = T_2$ but W_1 is much greater than W_2.)

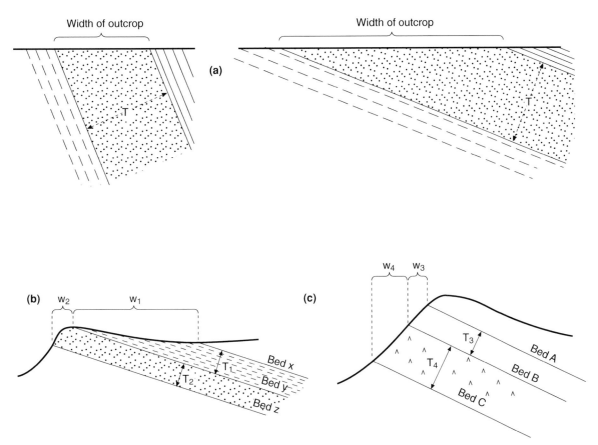

Fig. 3.8 (a) Sections showing the different widths of outcrop produced by a bed of the same thickness (T) with high dip and low dip. (b) Beds of the same thickness (T1 = T2) outcropping on differing slope. (c) Beds of different thickness (T4 = T3 × 2) outcropping on a uniform slope.

It will be noted that in the case of horizontal strata the geological boundaries are parallel to the topographic contours. In dipping strata the geological boundaries cross the topographic contours, and with irregular topography the steeper the dip, the straighter the outcrops. In the limiting case, that of vertical strata, outcrops are straight and unrelated to the topography.

Inliers and outliers

An outcrop of a bed entirely surrounded by outcrops of younger beds is called an inlier. An outcrop of a bed entirely surrounded by older beds (and so separated from the main outcrop) is called an outlier. In Map 5 these features are the product of erosion on structurally simple strata

and are called 'erosional' inliers and outliers. (See p. 71 for details of other inliers and outliers.)

Vertical exaggeration

It is very difficult to make the vertical scale equal to the horizontal scale for a one-inch-to-the-mile map, which is, of course, 1″ = 5280 feet. It is necessary to introduce a vertical exaggeration, which should be kept to the minimum practicable. No hard and fast rule can be made. The British Geological Survey commonly employs a vertical exaggeration (VE) of two or three, but sections provided on BGS maps range from true scale to a vertical exaggeration of as much as ×10 where this is necessary. The immensely useful 'Geological Highway Map' series of the USA (published by

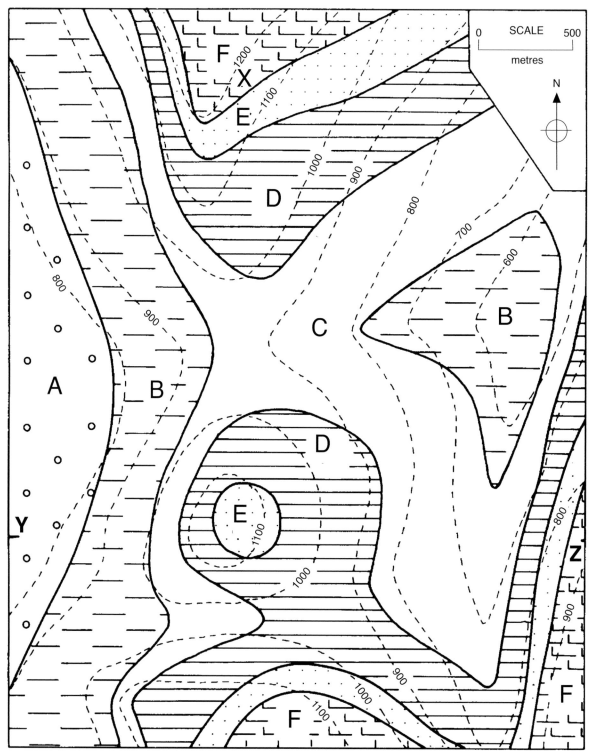

Map 5 Draw structure contours on the geological boundaries. Give the gradient of the beds (dip). Draw a section along the east–west line Y–Z. Calculate the thicknesses of beds B, C, D and E. Indicate on the map an inlier and an outlier (see p. 19).

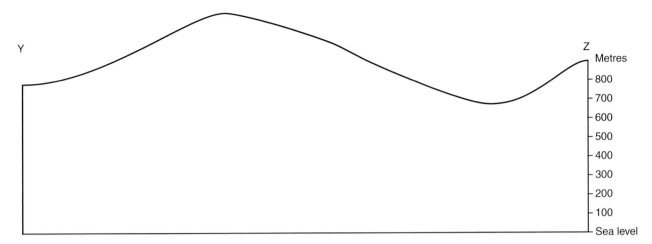

Fig. 3.9 Topographic profile of section Y–Z for Map 5.

and obtainable from the AAPG, Tulsa, Oklahoma) illustrates the effects of an overlarge VE of ×20. With one-inch-to-the-mile maps, a vertical scale of 1″ = 1000 feet is particularly convenient and the VE at approximately ×5¼ is generally acceptable.

It is essential to realise that where the section is drawn with a vertical exaggeration of three, for example, the tangent of the angle of dip must be multiplied by three to find the angle of dip appropriate to this section. (It is not the angle of dip that is multiplied by three.) A dip of 20° should be shown on a section with a VE of ×3 as a dip of 48°. Look up the tangents and check that this is correct.

Exercises using geological survey maps

1. **Market Rasen: 1:50,000 (Sheet 102) Solid & Drift edition, 1999.** This recent map shows simply dipping strata with a very small (or low) easterly dip. The outcrops run approximately north–south. The hardest bed, most resistant to erosion, the Marlstone Rock formation, forms a west-facing scarp. It and overlying beds, the Whitby Mudstone Formation and Grantham Formation, have almost straight boundaries to their outcrops. Further to the east, where streams have cut into the dip slope of the cuesta, the outcrops 'V' down the valleys. See Fig. 3.9.

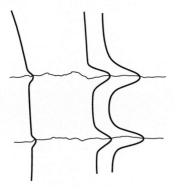

Fig. 3.9 Sketch map shows the characteristic 'V-shaped outcrops' in which the 'V' always points in the direction of dip.

2. **Henley-on-Thames: 1:50,000 (Sheet 254) Solid & Drift edition, 1980.** Examine the map and section, noting the relationship of topography to geology. (The Chalk is one of the most resistant formations in South East England, forming the high ground of the Chilterns, Downs, etc.) The presence of Drift Deposits, especially widespread in the Thames valley, makes it a bit more difficult to find the numerous outliers to the west of the main escarpment but, fortunately, the positions of geological boundaries beneath the cover of drift are clearly indicated by broken lines. Try to find the outliers on the map. They fall into two groups: (1) Cenomanian Lower Chalk (with Glauconite Marl [M] forming the lowest bed) rests on upper Greensand [UGS] and (2)

Turonian Middle Chalk (with Melbourne Rock [MbR] forming the lowest bed) rests on the Lower Chalk.

A map of structure contours for the base of the Lower Chalk and for the base of the Upper Chalk is included on the Henley-on-Thames sheet. It shows that these strata are gently folded (or flexured) and shows that the pattern of folding in both formations is similar (as one would expect).

3. **Aylesbury: 1" (Sheet 238) Solid & Drift edition, 1997.** Excluding the rather extensive Pleistocene and Recent deposits (which form a quite thin superficial cover), the oldest strata are to be found in the north-west of the area with successively younger beds to the south-east. This gives the direction of dip. The amount of dip can best be determined by ensuring that beds are made the appropriate thickness on the section, e.g. Lower Chalk and Middle Chalk should each measure approximately 200 feet if the section has been correctly drawn. Draw a section along a line in a north-west–south-east direction across the map to illustrate the structure of the area.

Note on the Aylesbury and Henley-on-Thames sheets. When dealing with an area the size of one of these maps, it is found that strata are not uniformly dipping (as is the case on the simpler problem maps) but may be slightly flexured. Structure contours – if they could be drawn – would not be quite straight, nor precisely parallel. The beds cannot be inserted on a section by constructing structure contours: they must be 'fitted' to the outcrop widths and drawn at their correct thicknesses, using information given in the stratigraphic column in the margin of the map.

Notes on BGS maps. The British Geological Survey produces a much wider variety of maps than formerly, both in range of scales employed and in the kind of information included.
   Most of England and parts of Wales, together with about half of Scotland, have now been covered by maps published on the 1:50,000 scale. Some areas are covered by still available and most useful one-inch-to-the-mile sheets. Both Solid geology editions and Drift editions have been produced for some areas. For the purposes of investigating geological structures the Solid editions are the more useful. However, Drift editions, showing the superficial deposits, are vital to the engineering geologist concerned with planning motorways, dams and foundations. In areas where drift deposits are not extensive, a single edition combining Solid and Drift is usually published.

4

Three-point problems

If the height of a bed is known at three or more points (not in a straight line), it is possible to find the direction of strike and to calculate the dip of the bed, provided dip is uniform. This principle has many applications to mining, opencast and borehole problems encountered by applied geologists and engineers but this chapter deals only with the fundamental principle and includes a few simple problem maps.

The height of a bed may be known at points where it outcrops or its height may be calculated from its known depth in boreholes or mine shafts. If the height is known at three points (or more) only one possible solution exists as to the direction and amount of dip, and this can be simply calculated.

Construction of structure contours

Note: Since Map 6 portrays a coal seam, and an average seam is of the order of 2 m or less in thickness, on the scale of 2.5 cm = 500 m its thickness is such that it can satisfactorily be represented by a single line on the map. There is no need to attempt to draw structure contours for the top and the base of the seam – on this scale they are essentially identical.

Observe the height of the seam at points A, B and C where it outcrops. Join with a straight line the highest point on the coal seam, C (600 m), to the lowest point on the seam, A (200 m). Divide the line A–C into four equal parts (since 600 m – 200 m = 400 m). As the slope of the seam is constant we can find a point on A–C where the seam is at a height of 400 m (the mid-point). We also know that the seam is at a height of 400 m at point B. A straight line drawn through these two points is the 400 m structure contour. On a simply dipping stratum such as this, all structure contours are parallel. Construct the 200 m structure contour through point A, and the 300 m, the 500 m and the 600 m structure contours – the latter through point C. Having now established both the direction and

the spacing of the structure contours, complete the pattern over the whole of the map.

Depth in boreholes

Relative to sea level, the height of the ground at the site of a borehole can be estimated from its proximity to contour lines and that of the coal seam at the same point on the map can be calculated from the structure contours. Quite simply, the difference in height between the ground surface and the seam is the depth to which the borehole must be drilled to reach the seam.

Notes on Map 6. At what depth would the coal seam be encountered in a borehole situated at point D? Since the borehole at D is sited on the 400 m topographic contour and (as you will see when you have constructed the structure contours for the coal seam) the 200 m structure contour passes through point D, the depth of the coal is equal to the height of the ground minus the height of the coal (400 m – 200 m). The depth to the coal is therefore 200 m.

Wherever a topographic contour intersects a structure contour we know both the height of the ground and the height of the coal bed. By simple arithmetic we can find the depth to the coal. If we join all the points where the seam is 200 m below ground level (as is the case in Borehole D) we are constructing a line called the 200 m isopachyte. (An isopachyte is defined on p. 71 where more complicated examples are to be found.) On Map 6, draw in the isopachytes for 100 m and 200 m. They will tend to be roughly parallel to the outcrop (which is in effect the 'zero isopachyte'). If you find this problem difficult, return to it after you have worked out Map 20.

Insertion of outcrops

The structure contours were drawn by ascertaining the height of the coal seam where it outcropped on contour lines. Wherever the seam – defined by its structure contours – is at the same height as the ground surface – defined by topographic contour lines – it will outcrop. We can find on the map a number of intersections at which structure contours and topographic contours are of the same

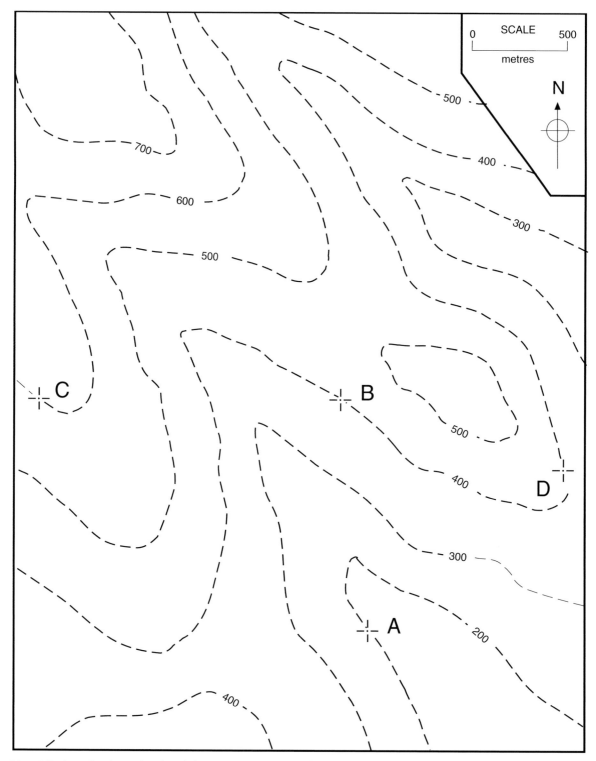

Map 6 Deduce the dip and strike of the coal seam that is seen to outcrop at points A, B and C. At what depth would the seam be encountered in a borehole sunk at point D? Complete the outcrops of the seam (see Fig. 4.1). Would a seam 200 m vertically below this one also outcrop within the area of the map? If yes, sketch in its outcrop. Contours in metres.

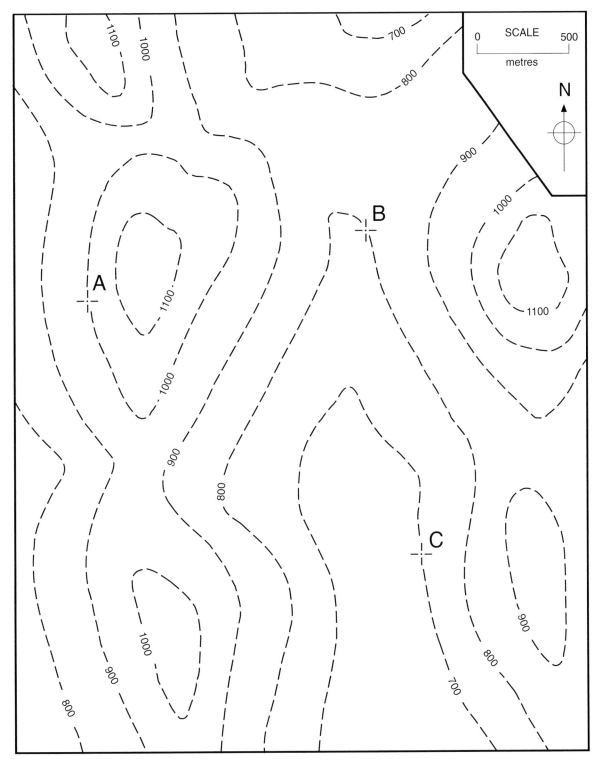

Map 7 Borehole A passes through a coal seam at a depth of 50 m and reaches a lower seam at a depth of 450 m. Boreholes B and C reach the lower seam at depths of 150 m and 250 m, respectively. Having determined the dip and strike, map in the outcrops of the two seams (assume that the seams have a constant vertical separation of 400 m). Indicate the areas where the upper seam is at a depth of less than 150 m below the ground surface. It is necessary to first calculate the height (relative to sea level) of the lower coal seam at each of the points A, B and C where boreholes are sited.

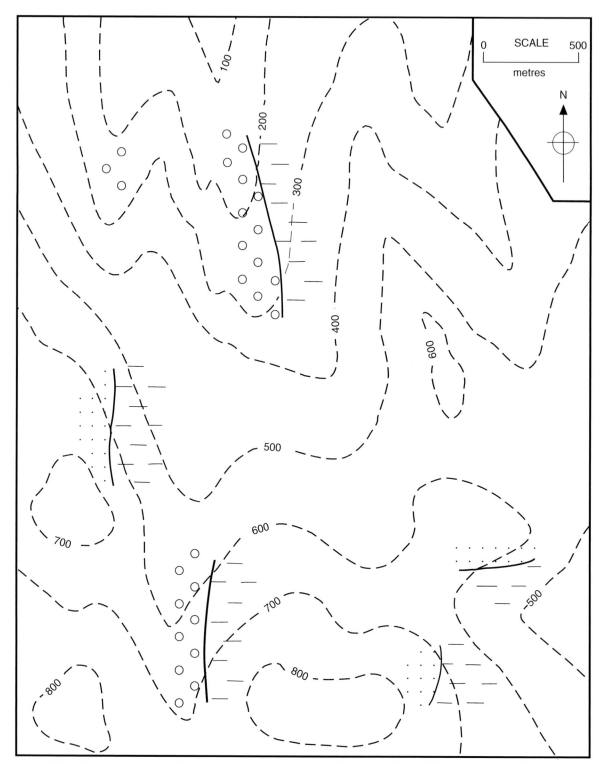

Map 8 Three beds outcrop – conglomerate, sandstone and shale. Complete the geological boundaries between these beds, assuming that the beds all have the same dip. Indicate on the map an inlier and an outlier.

height: the outcrop of the seam must pass through all these points.

Further, these points cannot be joined by straight lines. We must bear in mind that where the seam lies between two structure contours, e.g. the 300 m and 400 m, it can outcrop only where the ground is also at a height of between 300 m and 400 m, i.e. between the 300 m and 400 m topographic contours (Fig. 4.1). The outcrop of a geological boundary surface cannot, on a map, cross a structure contour or a topographic contour line except where they intersect at the same height.

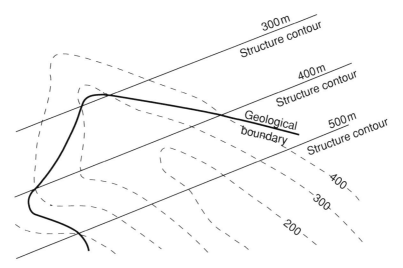

Fig. 4.1 The insertion of a geological boundary on a map with topographic contours and structure contours.

5

Unconformities

In terms of geological history, an unconformity represents a period of time during which strata are not laid down. During this period, strata already formed may be uplifted and tilted by earth movements which also terminate sedimentation. The uplifted strata, coming under the effects of sub-aerial weathering and erosion, are 'worn down' to a greater or lesser extent before subsidence causes the renewal of sedimentation and the formation of further strata. As a result we find, in the field, one set of strata resting on the eroded surface of an older set of beds.

Many of the features that characterise unconformities cannot be deduced from map evidence alone. For example, only in the field can one observe an actual erosion surface including the presence of palaeosols and hardgrounds, which represent periods of non-deposition but without major tectonic upheaval and tilting. An example is shown in Fig. 5.1.

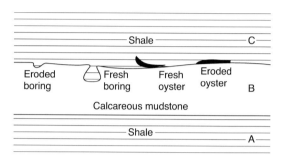

Fig. 5.1 Hardground forming a parallel unconformity.

Here, a lime-rich sediment (B) has been lithified on the sea floor and has subsequently been colonised by two generations of oysters and boring bivalves (piddocks). The fact that the first generation has suffered erosion is evidence that the sea bed has been an active erosion zone prior to the deposition of the next sediment (C). This is a

Plate 5 Unconformity, Dolyhir, Kington, Herefordshire. Beneath the unconformity there are dipping beds of Precambrian sedimentary rocks. Above the unconformity there are nearly horizontal beds of Silurian shaly limestone.

parallel unconformity as it involves no tilting of strata.

Derived fragments of the older strata may be re-deposited in the post-unconformity strata (sometimes as a basal conglomerate) and the analogous but much rarer phenomenon of derived fossils may be seen. However, the evidence of a gap in the stratigraphic succession should be indicated in the stratigraphic column provided on a map and some of the above features may be referred to in the data given.

On a map the main evidence of unconformity is a difference in the dip and strike directions in the pre- and post-unconformity strata. Earth movements which, by uplift, terminated sedimentation and subsequent sinking (which in turn permitted resumed sedimentation) frequently resulted in differences in dip and strike.

An exception to this can be found on the margins of the London Basin, where the Lower Tertiary beds rest unconformably on the Chalk with little difference in strike or dip, yet, by comparison with successions of strata on the other side of the Channel, we know that this is a major unconformity representing a long period of time since, in Britain, the uppermost stage of the Chalk and the lowest two stages of the Tertiary are absent.

In some cases, an unconformity represents a time interval of such length that the older strata are subject not only to tilting and uplift but also to fracture (faulting) and subsequent intrusion. In the cliff face shown in Fig. 5.2, the dyke has been intruded along a fault, as evidenced by the displaced strata (A–E) on each side. A much later transgression of the sea has deposited Bed F on top of the former landscape along an angular unconformity. Subsequently, a further hiatus between the deposition of Beds G and H has produced a second but younger unconformity. In this particular case, the major differences in dip and direction above and below the unconformity would be clearly seen on a geological map.

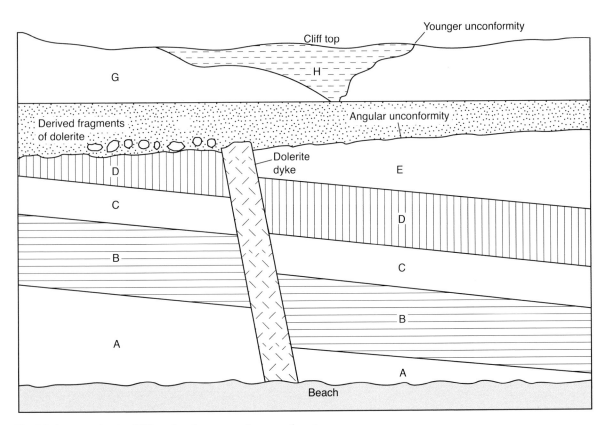

Fig. 5.2 An example of a cliff face showing an angular unconformity.

Plate 6 Unconformity between dipping grey Carboniferous Limestone and horizontally bedded, light brown, oolitic limestone of Jurassic age. First described by De la Beche (1846). Vallis Vale (Grid reference ST 756491), Eastern Mendip Hills, Somerset.

Overstep

Usually, the lowest bed of the younger series of strata, having a quite different dip and strike from that of the older strata, rests on beds of different age (see Plate 5). This feature (Fig. 5.3) is called overstep; bed X is said to overstep beds A, B, C, etc.

If the older strata were tilted before erosion took place, they meet the plane of unconformity at an angle, and there is said to be an 'angular unconformity' (Fig. 5.3).

Overlap

As subsidence continues and the sea, for example, spreads further on to the old land area, successive beds are laid down, and they may be of greater geographical extent, so that a particular bed spreads beyond, or overlaps, the preceding bed. This feature (Fig. 5.4) may accompany an unconformity with or without overstep. Bed Y overlaps Bed X. (The converse effect, that of successive beds being laid down over a progressively contracting area of deposition, due to gradual uplift, is known

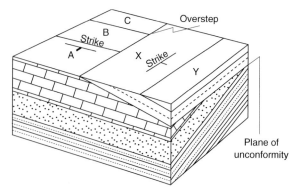

Fig. 5.3 Block diagram of an angular unconformity.

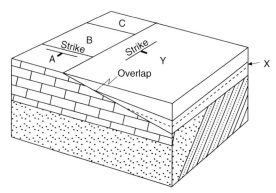

Fig. 5.4 Block diagram of an unconformity with overlap.

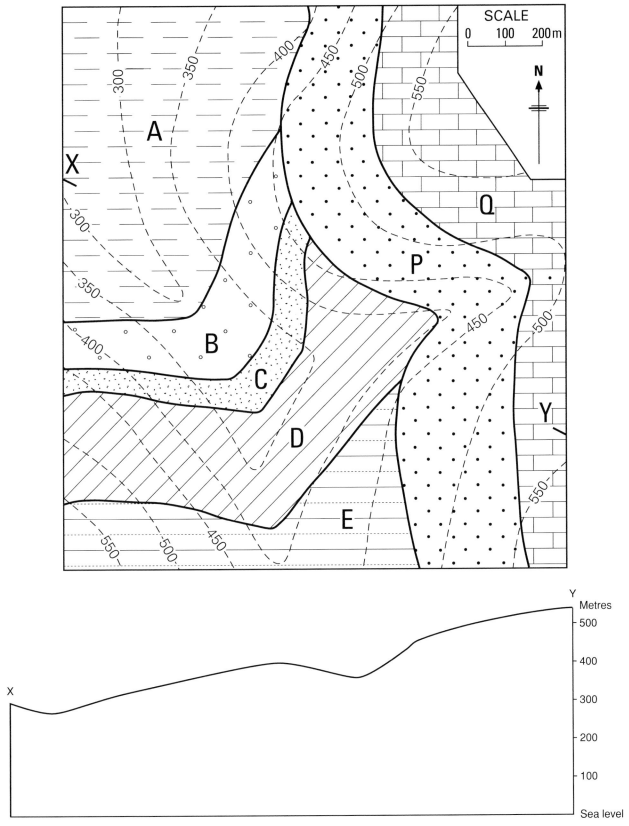

Map 9 Indicate on the map the outcrop of the plane of unconformity. Work out the dip and strike of the series of beds A to E and of beds P and Q. Note the difference in the strike direction of the two series, the most significant indication of unconformity from map evidence. Draw a section along the line X–Y on the profile provided.

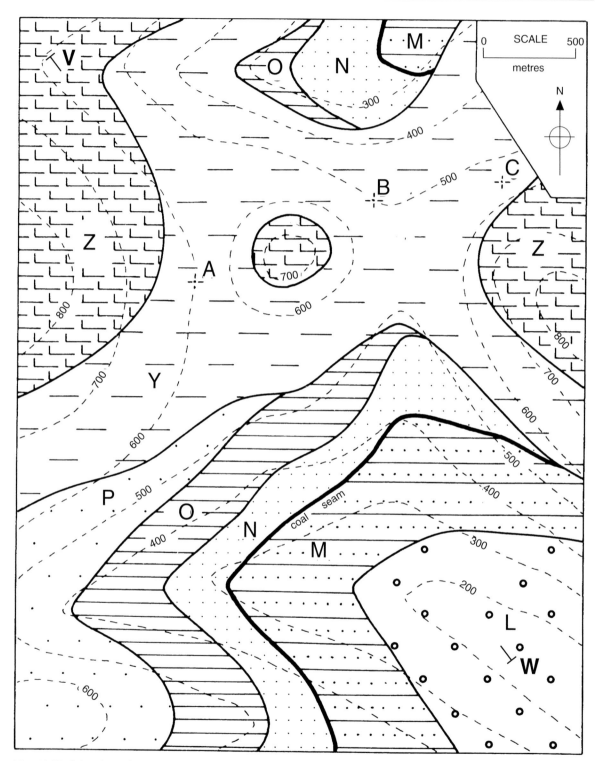

Map 10 Find the plane of unconformity. Deduce the direction and amount of dip of the two series of beds. Draw a section along the line from V near the north-west corner of the map to W near the south-east corner. Would the coal seam be encountered in boreholes situated at points A, B and C? If the coal is present, calculate its depth below the ground surface; if it is absent, suggest an explanation for its absence. Indicate the position of the coal seam beneath Bed Y.

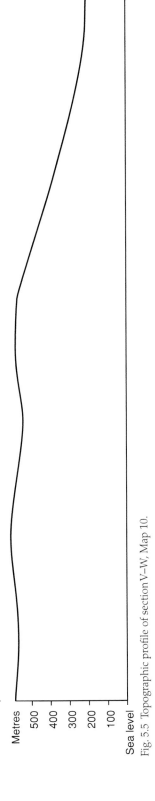

Fig. 5.5 Topographic profile of section V–W, Map 10.

Note on Map 10. The sub-unconformity position of the coal seam may be obtained by joining with a straight line the two points where the coal seam meets the base of bed Y (which is the plane of unconformity). To confirm that this line is correct, note where the 300 m structure contour for the coal seam intersects the 300 m structure contour on the base of bed Y. Also note where the 400 m structure contour for the coal intersects the 400 m structure contour on the base of Y. Both these points should lie on the line that you have drawn.

as off-lap. Such a feature is rarely deducible from a geological map and will be discussed no further.)

Two principal types of unconformity are recognisable:

1. planes of marine erosion, readily definable by structure contours;
2. buried landscapes. Sub-aerial erosion did not produce a peneplain and the post-unconformity sediments were deposited on a very irregular surface of hills and valleys, gradually burying these features.

Problem maps 9, 10 and 20 are exercises based on type (1). Usually, it is necessary to look at a broader regional geology to see the effects of buried landscape. See below the note on the Assynt one-inch geological sheet.

Sub-unconformity outcrops

An unconformity represents a period of erosion. Tens or hundreds of metres of strata may be removed. How can we examine this phenomenon and deduce what remains of the older strata since they are now covered by post-unconformity strata? We could, of course, in practice remove all post-unconformity strata with a bulldozer and other earth-moving equipment to reveal the plane of unconformity. We can, by a study of the structure contours, deduce from a map just what we should expect to see if that were possible.

What we seek are the 'outcrops' of the older strata on the plane of unconformity. This plane can be defined by its structure contours. If we now take the structure contours drawn on the geological boundaries of the older set of strata and note where they intersect, not ground contours, but the structure contours of the plane of unconformity, we can plot – where the two sets of structure contours intersect at the same height – a number of points which will define the sub-unconformity outcrops. See Note on Map 10. The topic will be revised in Chapter 10 and Maps 12 and 20 give further practical exercises.

Exercises using geological survey maps

1. **Sheffield: 1:50,000 (Sheet 100) Solid & Drift edition, 1974.** There is a very conspicuous unconformity between the Permo-Triassic Beds and the underlying Westphalian Coal Measures. Figure 5.6 shows the succession.

   On the map note that the Magnesian Limestone overlaps the lower Permian Marl, which in turn overlaps the Permian Basal Sands.

   The outcrop of unconformity is somewhat complicated by the presence of post-Permo-Triassic faults. Note the evidence of unconformity is that the Permo-Triassic Beds overstep on to Westphalian Beds of different horizons (ages) and they overstep faulted Westphalian where there are faults of pre-Permo-Triassic age. Look for examples of these, one in the south of the map where faulted Westphalian is overstepped by Basal Sands, and another near Grid Line 85 where faulted Westphalian is overstepped by Magnesian Limestone.

2. **Assynt: 1″ Geological Survey special sheet, Solid & Drift edition, 1965.** Examine the western part of the map to find the unconformities at the base of the Torridonian and the base of the Cambrian. Draw a section along the north–south grid line 22 to show these unconformities. From your knowledge of the conditions under which the Torridonian and Lower Cambrian were deposited, can you explain how and why these unconformities differ?

3. **Shrewsbury: 1″ (Map No. 152), Solid edition, 1978.** Examine the map carefully. How many of the unconformities indicated in the geological column in the margins are deducible from the map evidence?

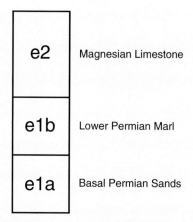

Fig. 5.6 Permian succession in the Sheffield area.

6

Folding

We have seen that strata are frequently inclined (or dipping). On examining the strata over a wider area it is found that the inclination is not constant and, as a rule, the inclined strata are part of a much greater structure. For example, the Chalk of the South Downs dips generally southwards towards the Channel – as can be determined by examination of the 1:50,000 Geological Survey map of Brighton & Worthing (Sheet No. 318/333). We know, however, that in the North Downs the Chalk dips to the north (passing beneath the London Basin); the inclined strata of the South and North Downs are really parts of a great structure that arched up the rocks including the Chalk over the Wealden area (as illustrated in Fig. 6.4).

The process of folding within a rock succession represents a response to the application of stress, which initiates deformation. The actual physical process whereby apparently solid and unmovable bodies of rock become deformed into a series of complex folds (Plate 7), as seen near Lulworth Cove in Dorset, is perhaps initially difficult to comprehend. However, these processes take place deep within the compression zones of mountain belts, squeezed between advancing tectonic plates, where temperatures and pressures are much higher and, given enough time, such deformation as seen at Lulworth is indeed possible.

An analogy of the folding process is shown in Fig. 6.1. Here a rubber tube, representing a solid rod of rock, is attached firmly to a wall. On pulling, the tube extends elastically. Provided its elastic limit is

Plate 7 The Lulworth Crumple. Complex folding and minor faulting within steeply dipping Middle & Upper Purbeck Beds. The folding of the more competent Portland Limestone (seen to the right) has led to gravitational collapse structures and brecciation within the less competent Purbeck Beds.

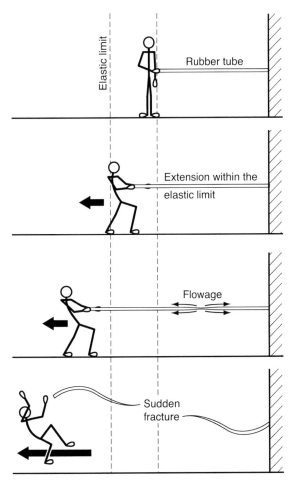

Fig. 6.1 Analogy of the folding and faulting processes.

not exceeded, it will return to its former length and not show any deformation. However, if its elastic limit is exceeded, the tube will change shape without breaking – that is, it will begin to thin at a weak point and ductile deformation will have occurred. Eventually, with continued stretching, the tube will break as brittle deformation is initiated. The longer the time period between the first observable deformation (i.e. thinning) and eventual breakage, the more ductile is the rock material. The speed of deformation is also important as the slower the application of stress, the more likely the rock will deform in a ductile fashion. The slow movement of tectonic plates provides an ideal scenario for this to occur.

In geological terms, the ductile flow phase is therefore represented by *folding* processes, while the brittle phase is indicative of *faulting*. We will be returning to faulting in Chapter 8.

An important result from our studies of rock deformation is that folding of rocks must precede the faulting of the same rock succession. The faulting can itself then provide fissures and fractures to be exploited by igneous intrusions and in terms of analysing a series of geological events, the order Folding → Faulting → Intrusion is sacrosanct and geological landscapes are made up of a series of these episodes, often separated by unconformities.

Another important factor is that of the composition of the rock types involved within the deformation. Under stress, rock types such as mudstones, shales, volcanic ashes and evaporites (e.g. halite or gypsum) will initially behave in a ductile fashion. Under the same stress regime, in rocks such as limestone and sandstone, the ductile-to-brittle transition will be much shorter and fracturing, usually in the form of tension joints, will occur around the fold noses as they develop.

Figure 6.2 shows the development of a tightly folded arch within a series of limestones and shales. The limestones show clear joints and are said to be competent. By contrast, the shales show flowage into a series of complex parasitic folds, which results in the beds being thinner on the limbs of the structure and thicker in the nose of the fold. The shales are said to be incompetent.

The complex patterns shown by the parasitic folding are important when dealing with plunging folds and areas of high structural complexity. This will be covered more fully in Chapter 13.

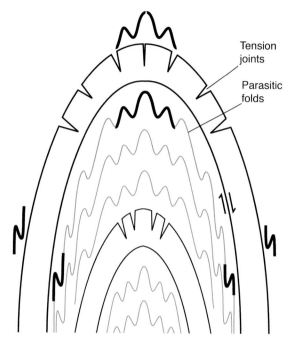

Fig. 6.2 The differing responses of a series of competent and incompetent beds to folding.

Anticlines and synclines

Where the beds are bent upwards into an arch the structure is called an anticline (*anti* = opposite; *clino* = slope; the beds dip away from each other on opposite sides of the arch-like structure). Where the beds are bowed downwards the

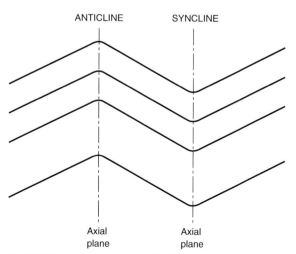

Fig. 6.3 Diagrammatic section of folded strata. Compare with Plate 8.

structure is called a syncline (*syn* = together; clino = slope; beds dip inwards towards each other) (Fig. 6.3). In the simplest case the beds on each side of a fold structure, i.e. the limbs of the fold, have the same amount of dip and the fold is symmetrical. In this case a plane bisecting the fold, called the axial plane, is vertical. The fold is called an upright fold whenever the axial plane is vertical or steeply dipping. (Where axial planes have a low dip or are nearly horizontal, as in Fig. 13.1(a), folds are called flat folds.)

anticline the oldest bed outcrops in the centre of the structure and, as we move outwards, successively younger beds are found to outcrop (Fig. 6.4). In an eroded syncline, conversely, the youngest bed outcrops at the centre of the structure, with successively older beds outcropping to either side (Fig. 6.5).

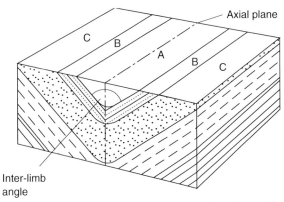

Fig. 6.5 Block diagram of a symmetrical syncline (an upright fold).

Plate 8 Anticline, Saundersfoot, Pembrokeshire. This anticline is asymmetrical because its northern limb is steeper than the southern one. The rocks are Upper Carboniferous Coal Measures sandstones.

The effect of erosion on folded strata is to produce outcrops such that the succession of beds of one limb is repeated, though of course in the reverse order, in the other limb. In an eroded

Plate 9 A broad anticlinal fold within a series of greywacke sandstones and subsidiary slaty horizons. A strong axial plane cleavage and parasitic folds are also visible. South Stack, Anglesey, NW Wales.

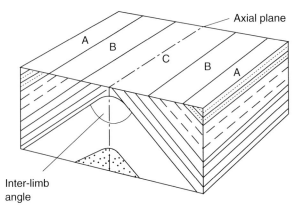

Fig. 6.4 Block diagram of a symmetrical anticline (an upright fold).

Asymmetrical folds

In many cases the stresses in the earth's crust producing folding are such that the folds are not symmetrical like those described above. If the beds of one limb of a fold dip more steeply than the beds of the other limb, then the fold is asymmetrical (see Plate 8). The differences in dip of the beds of the two limbs will be reflected in the widths of their outcrops, which will be narrower in the case of the limb with the steeper dip (Fig. 6.6). (See also p. 19 and Fig. 3.8.) Now, the axial plane bisecting the fold is no longer vertical but is inclined and the fold is called an inclined fold.

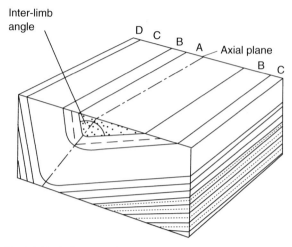

Fig. 6.6 Block diagram of an asymmetrical syncline (an inclined fold).

Problem 1

Using the map on p. 39, which is part of the BGS Brighton & Worthing 1:50,000 Geological sheet, answer the following questions. This particular map is a Solid & Drift edition which shows both solid and drift geology (Quaternary) on the same sheet. This is possible because the considerable variety of Quaternary deposits are not extensive and do not obscure much of the solid geology. In fact, the Quaternary deposits are helpful in interpreting the drainage (of rivers and streams).

1. Deduce the main features of the topography of the area, remembering that in the south and east of England the hardest and most resistant rocks are the limestones of Upper Cretaceous age known as the Chalk.

2. The chalk ridge is a steep north-facing escarpment. Further to the north the land can be seen to be generally between 20 m and 30 m in height. The geology of this area is particularly closely related to the topography. Check that this is so and summarise the relationship.

   In the middle of the map area the succession of beds is reversed as we travel northwards. We know that normally older strata dip beneath younger strata, therefore here they are not dipping southwards, as in the map generally, but northwards, forming

an anticline. Beyond, a syncline restores the dip direction to southerly.

3. Briefly describe the folds. Give the bearing of the axes. Note that the southern limb of the syncline is faulted. Which way does it downthrow?

   The drainage pattern, especially of the southern half of the area, is very informative. As the Wealden Anticline uplifted, rivers known as consequent streams developed on the dip slopes. The major river in this area, the River Adur, is an example, rising to the north of this map area in the Weald. It is an important river, still flowing southwards as the land rose and now cutting through the South Downs and entering the sea at Shoreham-by Sea. (Note that its entry to the sea has been deflected over 2 km to the east by eastwards longshore drift.)

Key

UMCK	Upper/Middle chalk
MbR	Melbourne rock
LCK	Lower chalk
UGS	Upper greensand
G	Gault clay

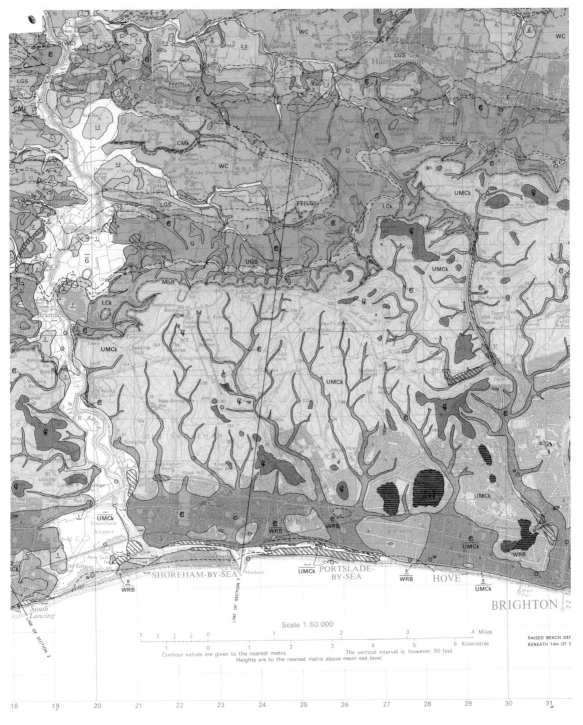

Fig. 6.7 **This map is reproduced by permission of the Geological Survey: IPR/132-29CT British Geological Survey © NERC. All rights reserved**
Geological Mapping, BGS © NERC, Ordnance Survey Topography © Crown Copyright. All rights reserved

Problem 1 – *continued*

Later, southward-draining streams have developed on the dip slope of the Chalk. The deposition of Head deposits in these valleys more clearly reveals on the map the dendritic pattern of their drainage.

To the north of the summit ridge of the Chalk, short streams flowing northwards have developed (obsequent streams) and they coalesce, flowing east or west before draining into the River Adur or another south-flowing river.

4. Figure 6.8 shows a typical trellised drainage pattern. Using this figure, indicate with a letter C the consequent streams and with CC the later consequent streams. Also indicate the subsequent streams (with a letter S) which have developed at right angles, and the obsequent streams that flow in the opposite direction to that of the consequent streams (with a letter O).

The youngest beds in this area, the Woolwich and Reading Beds and the London Clay, are shown as overlying the Chalk unconformably. This represents a considerable time interval as these beds are Palaeocene and Eocene in age but no overstep can be observed as they are obscured by widespread Head and Brickearth (Loess) deposits.

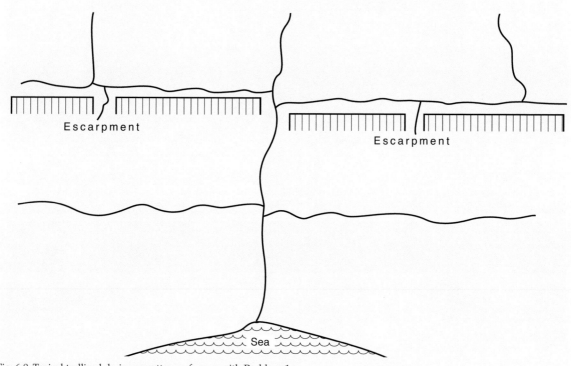

Fig. 6.8 Typical trellised drainage pattern – for use with Problem 1

Overfolds

Plate 10 Overturned fold within the Culm Measures (Carboniferous), Pinhoe, Exeter.

If the asymmetry of a fold is so great that both limbs dip in the same direction (though with different angles of dip), that is to say the steeply dipping limb of an asymmetrical fold has been pushed beyond the vertical so that it has a reversed – usually steep – dip, the fold is called an overfold (Fig. 6.9 and Plate 13). The strata of the limb with the reversed dip, it should be noted, are upside down, i.e. inverted.

Figures 6.5, 6.6, 6.9 and 6.11 show a progressive decrease in inter-limb angle and illustrate fold structures produced in response to increasing tectonic stress. Further terminology should be noted. Where the limbs of a fold dip at only a few degrees it is a gentle fold; with somewhat greater dip (Figs. 6.4 and 6.5) a fold is described as open; with steeper dipping limbs a fold is described as closed; and with parallel limbs (Fig. 6.8(a),(b)) the folding is isoclinal.

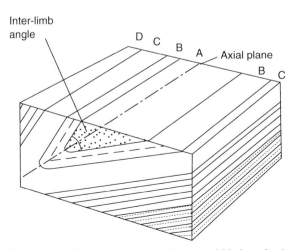

Fig. 6.9 Block diagram of an overfold (an overfolded syncline).

Plate 11 Overturned fold, Little Haven, Pembrokeshire. This anticline has its northern limb tilted through an angle greater than 90° so that the beds on this limb are overturned. The rocks are interbedded sandstones and shales of Carboniferous age. The shales are incompetent. Picture scale about 2 m across.

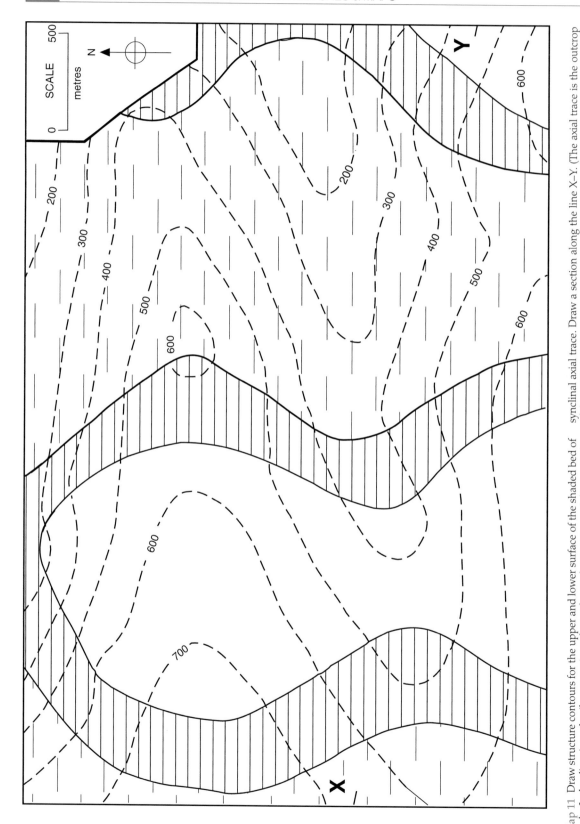

Map 11 Draw structure contours for the upper and lower surface of the shaded bed of shale. Is the direction of strike approximately north–south or east–west? (See p. 49 for help.) Indicate on the map the position of an anticlinal axial trace and the position of a synclinal axial trace. Draw a section along the line X–Y. (The axial trace is the outcrop of the axial plane.)

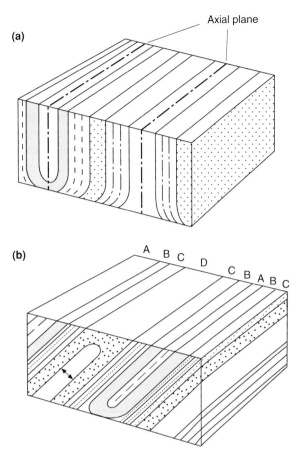

Fig. 6.11 Block diagrams of isoclinal folding: (a) upright folds; (b) overfolds.

Isoclinal folds

Isoclinal folds are a special case of overfolding in which the limbs of a fold both dip in the same direction at the same angle (isos = equal; clino = slope), as the term suggests (Fig. 6.11(b)). The axial planes of a series of such folds will also be approximately parallel over a small area, but over a larger area extending perhaps 40 kilometres (greater than that portrayed in a problem map) they may be seen to form a fan structure.

Way-up criteria

In any geological field investigation it is important to know in which order the sequence of rock units was formed. Way-upness, as it is termed, is therefore an inherent part of geological analysis and underpins the Law of Superposition defined by William Smith (Chapter 1).

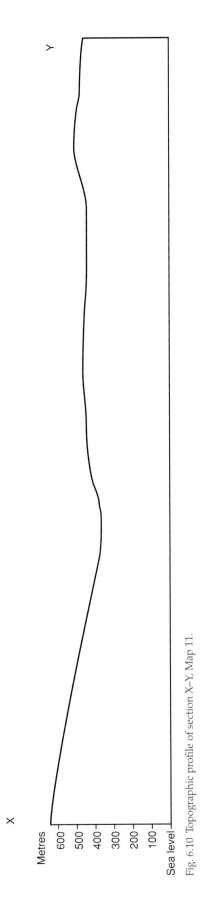

Fig. 6.10 Topographic profile of section X–Y, Map 11.

In this chapter both overfolds and isoclinal folds have been described. In each case, one of the fold limbs is overturned and therefore stratigraphically inverted. Within most sedimentary sequences there are small-scale structures that can prove the attitude of the sediments and which rocks are younger. This defines the direction of younging (see Fig. 6.12).

A small selection of sedimentological way-up criteria are considered here while further examples are found in igneous rocks (Chapter 11) and through cleavage bedding relationships (Chapter 13).

Cross-bedding

Cross-bedding (or cross-lamination on a small scale) is the general name given to structures where the rock displays angled lines sweeping down between the main bedding planes (Plates 10 and 11). These are formed when sand is deposited in a current of water or by the action of wind. Ripples or the larger dunes are the result of the continual motion of sand grains up the gentle back slope (stoss side), which eventually tumble down the steeper slope (lee side) (see Fig. 6.13). Both ripples and dunes therefore gradually migrate.

These structures are commonly truncated by the arrival of younger cross-bedded layers or sets. The result is a series of sloping layers that is concave upwards and thus clearly defines the direction towards younger strata.

Plate 13 Well-developed current bedding within the Old Red Sandstone, Cat's Back, South West Herefordshire. The beds are concave upwards and thus the sequence is right way up. (*Photograph by the late Peter Thomson*)

Graded bedding

This occurs with the deposition from a water current that begins as a strong flow but gradually loses power with distance travelled. River channel deposits with conglomerate bases grading up into finer sands in the upper layers is a common example of this process. Energetic turbidity flows moving from shallow continental shelves into deeper water also generate graded bedding.

Load casts

The common geological sequence of interbedded shales and sandstones in both freshwater and marine sequences produces a wide variety of way-up structures including load casts where the weight of the sand-grade sediments loads downwards into older and softer muds to produce the characteristic bulbous bases to the sandstone unit (see Fig. 6.12). This downward penetration causes the associated upward movement of the clays into flame structures.

Structures associated with animals and plants

The boring of worms or bivalves into rock surfaces, the upward growth and expansion of a colonial or solitary coral, and the penetration of roots downwards into the seat earth (Fig. 6.12) beneath a coal seam can also provide excellent evidence of the younging direction.

Plate 12 Ripple drift bedding within the Lower Cretaceous Woburn Sands, Leighton Buzzard, Bedfordshire. This structure shows the sedimentary sequence to be the right way up.

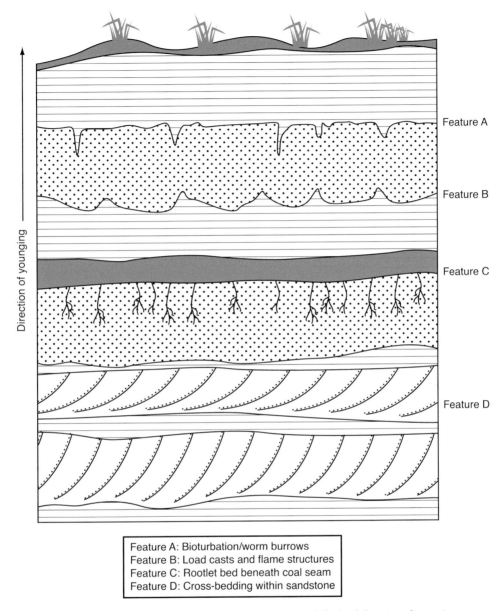

Fig. 6.12 Succession of sedimentary strata showing selected way-up criteria and the local direction of younging.

Feature A: Bioturbation/worm burrows
Feature B: Load casts and flame structures
Feature C: Rootlet bed beneath coal seam
Feature D: Cross-bedding within sandstone

Similar and concentric folding

When strata that are originally horizontal are folded it is clear that the higher beds of an anticline form a greater arc than the lower beds (and the converse applies in a syncline). Theoretically, at least two mechanisms are possible: the beds on the outside of a fold may be relatively stretched while those on the inside are compressed, or the

beds on the outside of a fold may slide over the surface of the inner beds (Fig. 6.14).

The way in which beds will react to stress depends upon their constituent materials (and the level in the crust at which the rocks lie). Competent rocks such as limestone and sandstone do not readily extend under tension or compress under compressive forces but give way by fracturing and buckling, while incompetent rocks such as shale

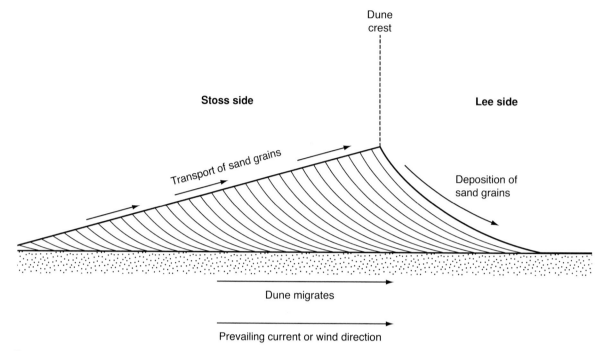

Fig. 6.13 A migrating dune showing the formation of cross-bedding.

or clay can be stretched or squeezed. Thus in an alternating sequence of sandstones and shales the sandstones will fracture and buckle, while the shales will squeeze into the available spaces.

Concentric folds

The beds of each fold are approximately concentric, i.e. successive beds are bent into arcs having the same centre of curvature. Beds retain their con-

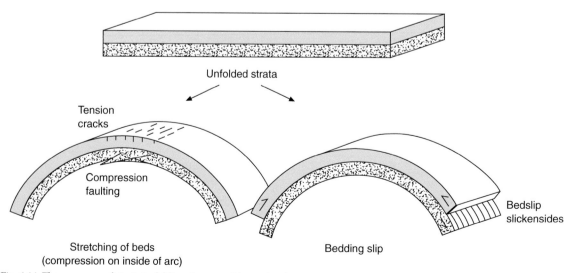

Fig. 6.14 The response of strata to folding: two possible mechanisms.

stituent thickness round the curves and there is little thinning or attenuation of beds in the limbs of the folds (Fig. 6.15(a)).

Straight-limbed folds also maintain the uniformity of thickness of the beds (except in the hinge of the fold) and folding takes place by slip along the bedding planes, as it does in the case of concentric folds. Although typically developed in thinly bedded rocks (such as the Culm Measures of North Devon), most of the problem maps in this book that illustrate folding have straight-limbed folds, since these provide a simple pattern of equally spaced structure contours on each limb of a fold.

Similar folds

The shape of successive bedding planes is essentially similar, hence the name (Fig. 6.15(b)). Thinning of the beds takes place in the limbs of the folds (and a strong axial plane cleavage is usually developed). This type of folding probably occurs when temperatures and pressures are high.

Two possible directions of strike

A structure contour is drawn by joining points at which a geological boundary surface (or bedding plane) is at the same height. By definition, this surface is at the same height along the whole length of that structure contour. Clearly, if we join points X and Y (Fig. 6.17) we are constructing a structure contour for the bedding plane shown, for not only are points X and Y at the same height but the bedding plane is at the same height along the line X–Y. If, however, we join the points W and X, although they are at the same height, we are not constructing a structure contour, as the bedding plane is not at the same height along the line W–X, it is folded downwards into a syncline. Thus, if we attempt to draw a structure contour pattern that proves to be incorrect, we should look for the correct direction approximately at right angles to our first attempt. It should also be noted that an attempt to visualise the structures must be made. For example, in Map 11 the valley sides provide, in essence, a section that suggests the synclinal structure, especially if the map is turned upside down and viewed from the north. Similarly, Map 13 reveals the essential nature of the structures by regarding the northern valley side as a section. To facilitate this, fold the map at right angles along the line of the valley bottom. Now regard the top half of the map as an approximate geological section.

What is the test of whether we have found the correct direction of strike? In these relatively simple maps the structure contours should be parallel and equally spaced (at least for each limb of a fold structure). Furthermore, calculations of true thicknesses of a bed at different points on the map should give the same value.

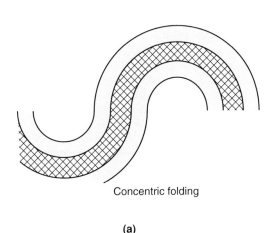

Concentric folding

(a)

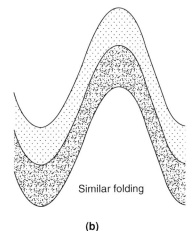

Similar folding

(b)

Fig. 6.15 The shape of concentric and similar folding seen in section.

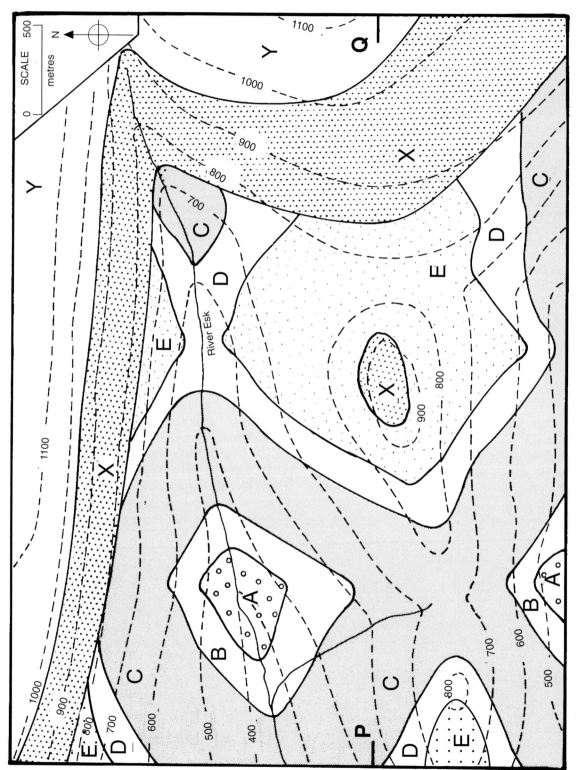

Map 12 Indicate on the map the outcrop of the plane of unconformity. Insert geology. Indicate on the map the extent of Bed E beneath the overlying strata. the axial traces of the folds. Draw a section along the line P–Q to illustrate the

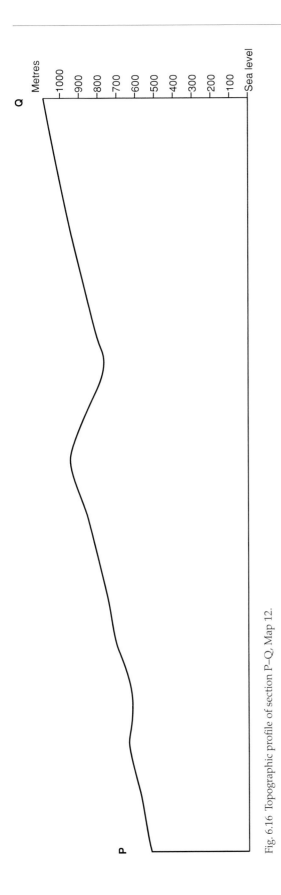

Fig. 6.16 Topographic profile of section P–Q, Map 12.

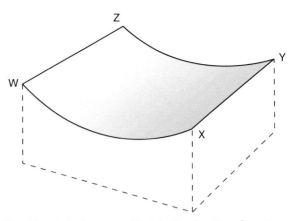

Fig. 6.17 Block diagram to illustrate the direction of structure contours in folded strata.

Remember that the topography is irrelevant to the solution of this part of the problem (see p. 33, 'Sub-unconformity outcrops'). The surface on which we are plotting the outcrop of the D/E boundary is the plane of unconformity, defined by the structure contours drawn on the base of Bed X. You will find it necessary to use intersections of structure contours which are now above ground level: although the strata in question have now been eroded away from these points, the points themselves remain valid as construction points.

Note on Map 12. You should find three areas where Bed E occurs underneath younger beds. The eastern extent of Bed E could be deduced by joining the three points where the D/E junction is cut by the base of Bed X. However, to confirm that this is correct it is necessary to use the intersection of the two sets of structure contours (on D/E and on the base of X). This method is the only way in which the western extent of Bed E can be defined.

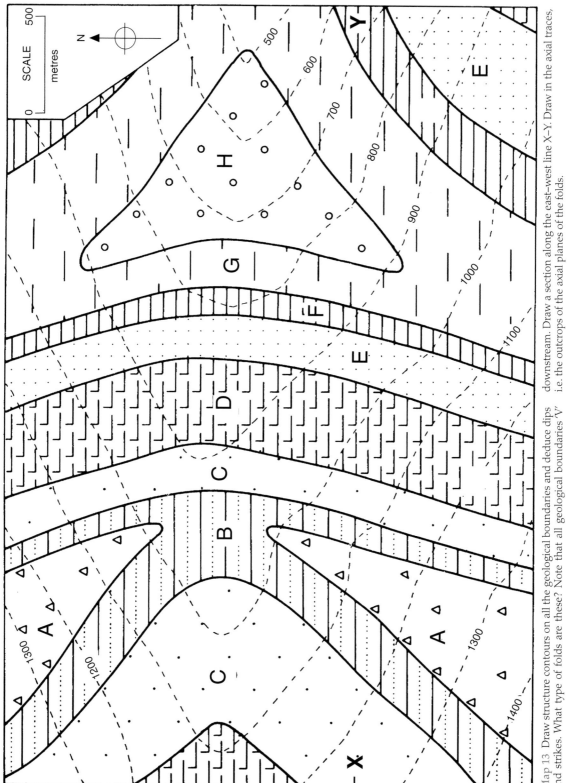

Map 13 Draw structure contours on all the geological boundaries and deduce dips and strikes. What type of folds are these? Note that all geological boundaries 'V' downstream. Draw a section along the east–west line X–Y. Draw in the axial traces, i.e. the outcrops of the axial planes of the folds.

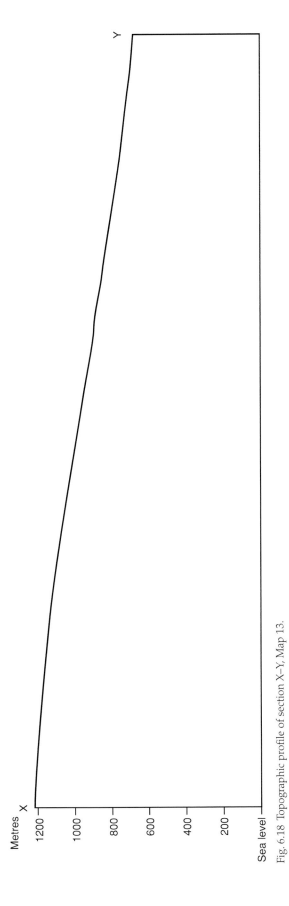

Fig. 6.18 Topographic profile of section X–Y, Map 13.

Lewes, 1:50,000, BGS Map (Sheet 319) Solid & Drift edition, 1979. Study the geology of this area on the southern limb of the Wealden anticline. Note the general structural trend, close to E–W, and the 'younging' of beds southwards. Study the relationship of topography to geology. Draw a section along the line of Section 1 (south of grid line 117). Make the vertical exaggeration ×4 (the same as the BGS section on this map).

Notes on the Lewes map. Here, as elsewhere in the south-east of England, the Chalk of the Upper Cretaceous age is a compact rock more resistant to erosion than the softer clays and sands of the pre-Chalk age. There is less information on the dip of strata than desirable, although dips are given to the north and south of the town of Lewes. Remember to allow for the vertical exaggeration by multiplying the gradient (tangent) of given dips by 4. Where dip information is less than adequate the beds must be fitted to the outcrops at an angle of dip that gives the correct thicknesses of strata shown in the stratigraphic column on the map.

Map solution without structure contours 1

This chapter introduces the interpretation of geological maps without recourse to drawing structure contours. It also acts as a revision of previous chapters.

Outcrop patterns depend on the geological structures and their intersection with the topography. In general, the topography is related to the underlying geology in a number of ways. In areas of relatively simple geological structures younger beds are found forming the hills while older strata outcrop and are exposed in the valleys. Hard rocks are more resistant to erosion than softer rocks, horizontal beds (and beds with low angle of dip) tend to produce steep escarpments, whereas steeply dipping beds generally give rise to more gentle slopes. Anticlines, their beds weakened by tension, are easily eroded. The result of these relationships is that structures such as unconformities, folds and faults, etc. can often be recognised from the typical outcrop patterns they produce. In many cases, therefore, it is possible to deduce the geological structure of an area from the outcrop patterns and other given information although, for one reason or another, it is not possible to construct structure contours (strike-lines).

Particularly in Geological Survey maps, much information is given on the map and in the margins. Dip of strata is shown at many places and anticlines and synclines are usually shown, often with the axial trace indicated. Faults are shown, often with an indication of the kind of fault, normal, reverse or wrench. The direction of downthrow is usually given and in some cases the amount. In areas of metamorphic geology information on foliation and lineation is given and shear zones may be indicated. In the margins of the map there will be symbols for these features.

The geological strata and igneous rocks may be arranged by their igneous and sedimentary classification (see key to Map 26) or they may be arranged by age groups (Fig. 7.1(a)). However, in areas of chiefly sedimentary rocks the strata are usually presented in the form of a stratigraphical column (Fig. 7.1(b)). This column will show the thicknesses of beds and their age. Lithologies, fossils and other characteristics may be described. Unconformities will be shown. Where given, this kind of information used in conjunction with outcrop patterns and other map evidence is invaluable.

The simplest cases to interpret are where geological structures outcrop on a flat surface. We find this when making large-scale geological maps of intertidal wavecut platforms and other erosion surfaces. While structure contours cannot be drawn, dip can usually be measured at a number of places. Exposure may be almost 100 per cent, if not obscured by loose blocks or other recent deposits (see Map 24).

At the other end of the scale broad plains or plateaux may be nearly flat and horizontal for tens of kilometres, even 100 km or more. Although they may be hundreds of metres above sea level (for example in West Texas or Arizona), it may be impossible to obtain the data needed to construct structure contours. The geological structures present must be deduced from outcrop patterns and information given in the legend of the map.

Typically, on maps drawn on a scale of 1:50,000 structure contours are seldom straight or parallel over the distances being covered. There is seldom sufficient evidence to permit structure contours to be drawn, although dip and strike will be given at a number of points where they have been measured and recorded by the field geologist. Additional useful data may be deduced from outcrop/contour intersections but, in general, on maps of this scale the interpretation is largely dependent on the outcrop patterns observed, together with a wealth of data given in the margins of the map (the legend). Problem Maps 16 and 23 are drawn on this scale.

What must you do to enable you to draw a section? Dip arrows show the inclination of the beds and values are given in degrees. To get the geology to fit the section and to allow for the doubled

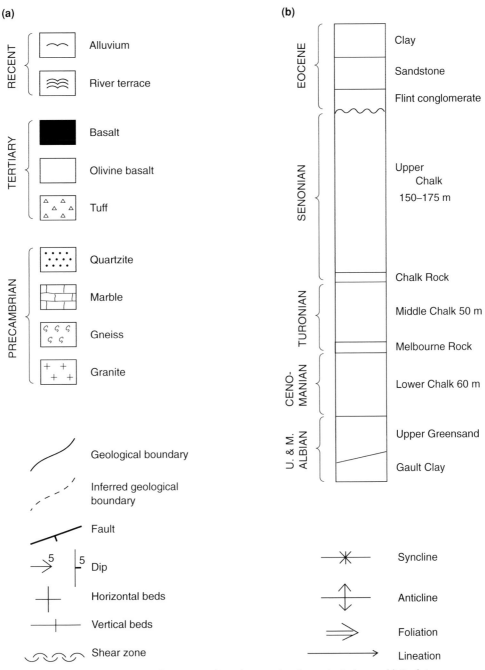

(a)

RECENT
{ Alluvium
  River terrace }

TERTIARY
{ Basalt
  Olivine basalt
  Tuff }

PRECAMBRIAN
{ Quartzite
  Marble
  Gneiss
  Granite }

Geological boundary

Inferred geological boundary

Fault

5 Dip

Horizontal beds

Vertical beds

Shear zone

(b)

EOCENE
{ Clay
  Sandstone
  Flint conglomerate }

SENONIAN
{ Upper Chalk 150–175 m }

TURONIAN
{ Chalk Rock
  Middle Chalk 50 m
  Melbourne Rock }

CENO-MANIAN
{ Lower Chalk 60 m }

U. & M. ALBIAN
{ Upper Greensand
  Gault Clay }

Syncline

Anticline

Foliation

Lineation

Fig. 7.1 Two ways in which information may be presented on the margin of a geological map. (a) Rock types grouped together according to their age, Precambrian, Tertiary, Recent (and in this case, coincidentally grouping together similar rock types (meta-morphic, igneous, sedimentary)). (b) A stratigraphic column. Below, some of the more common symbols encountered on a map, such as these, may be shown on the map margin. See also the key to Map 26.

vertical scale you must double the *gradient* (i.e. the tangent of the angle of dip). The shales and lime-stones dip to the north at 18°, a gradient of 1 in 3 = tangent 0.33. To use this in section drawing, double the gradient to 2 in 3 = tangent 0.66. (Fig. 7.4) (because of the vertical exaggeration), consult-ing the tangent table shows that 0.66 corresponds to an angle of 33°. Therefore you must increase the

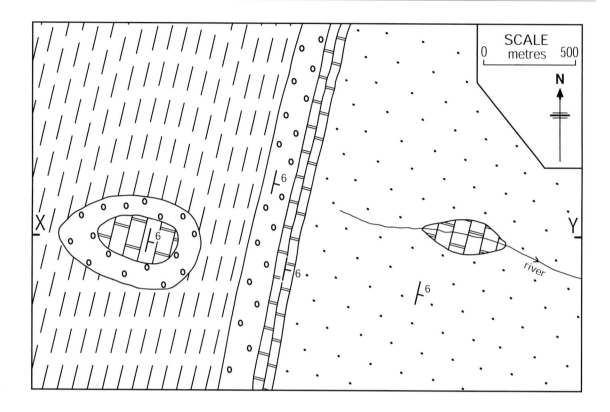

Map 14 Draw a section along the line X–Y on the topographic profile provided. Indicate on the map and on the section the inlier and the outlier. Reproduced from Fig. 14 of *Simple Geological*

Mapwork by WE Johnson (London, Edward Arnold, 1976) with kind permission of the author.

dip from 18° to 33°. With a low angle of dip, such as the 2° of the sandstone, it is sufficiently accurate to double the angle of dip of 2° to 4°. Use your tangent table to prove why this is so [tan 2° = 0.035, tan 4° = 0.07].

While the profile of this map could have been drawn with no vertical exaggeration without much difficulty, it illustrates how to deal with maps where the profile *must* be drawn with a vertical exaggeration in order to show the geology readily.

The advantages of not exaggerating the vertical scale, where this is practicable, are that it gives a correct picture of the structure, and thicknesses of beds can be measured on the section.

(A warning must be given that on published survey maps there sometimes appears to be a dip given that does not appear to fit in with the structural pattern deduced from the outcrop pattern and other information. Such data is not discarded since it is not clear whether this may be a local variation of significance.)

To recapitulate, consider outcrops on a near-horizontal plane surface. We saw in Chapter 3 that older beds dip under younger beds. This is a consequence of beds being deposited one on top of another with the oldest beds at the bottom and the youngest at the top (see Fig. 1.1 in Chapter 1). Subsequent tilting and erosion of the

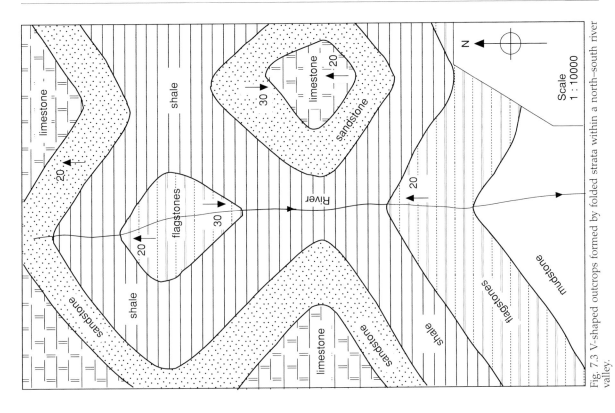

Fig. 7.3 V-shaped outcrops formed by folded strata within a north–south river valley.

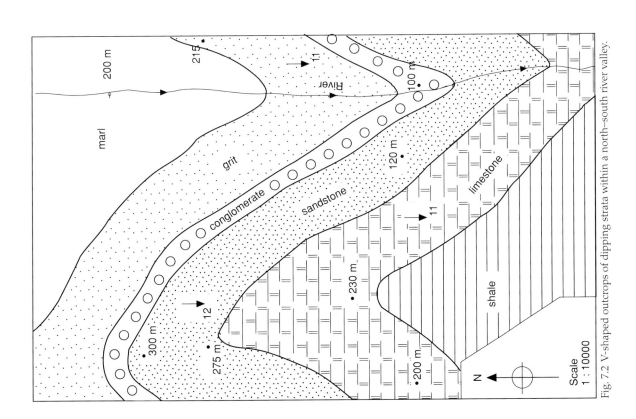

Fig. 7.2 V-shaped outcrops of dipping strata within a north–south river valley.

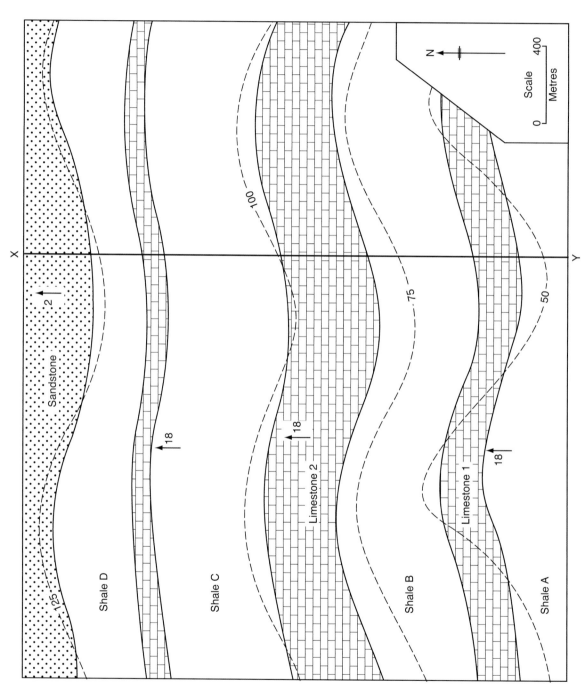

Map 15 The highest ground in the north of the area is capped by sand-stone that dips gently to the north. In the remainder of the area a series of limestone's and shale's outcrop dipping more steeply to the north.

Draw a section along the line X–Y on the profile provided to illustrate the geological structures. Note that the vertical scale of the profile is double that of the horizontal scale (the scale of the map, 1 cm = 200 m).

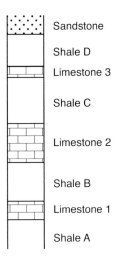

Sandstone

Shale D

Limestone 3

Shale C

Limestone 2

Shale B

Limestone 1

Shale A

Fig. 7.4 Key and Profile for use with Map 15.

X   Y

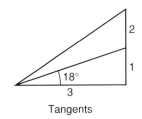

Tangents

Fig. 7.5 Tangents.

strata will result in the outcrop patterns seen on any map.

We also saw in Chapter 3 (pp. 18–19) that the steeper the dip, the narrower the outcrop of any bed. The extreme case occurs when beds are vertical – dipping at 90° – and the width of outcrop equals the actual thickness of the bed.

Where sequences of outcrops of beds are repeated in the reverse order, folding is usually responsible. The Order of Superposition (the relative ages of the strata) will indicate the direction of dip in the limb of a fold. It can readily be seen whether a fold is an anticline or a syncline

and whether overfolding is present (see Maps 11 and 13). The repetition of sequences of outcrops, or the suppression of parts of a sequence of outcrops is caused by faulting. This will be dealt with in Chapter 8. Unconformities are indicated in the field by erosion surfaces and such information may be given on a geological map. However, the essential evidence on the map is the difference in the direction of strike in the pre-unconformity sequence of strata and the direction of strike in the post-unconformity strata. (Only very rarely will they have a similar strike direction by chance.) Generally, an unconformable bed lies on several beds of the older series, since they may have been tilted or folded during the period of non-deposition.

Map 14 illustrates a typical outcrop pattern of dipping strata giving rise to an escarpment with a steep scarp slope and gentle dip slope. Imagine for a moment that there were no dip arrows on the map but that we knew the stratigraphical sequence – sandstone, limestone, conglomerate, marl. The oldest bed, the marl, outcrops in the west of the area while the youngest bed, the sandstone,

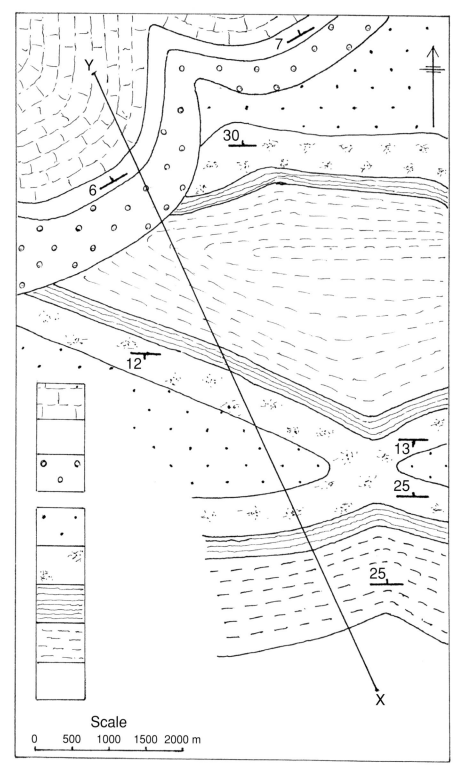

Map 16 Describe the structures shown on the map. Draw a section along the line X–Y on the profile provided. You will note that relief is slight except for a prominent scarp feature in the north-west of the area. Before commencing, check what the apparent dips will be in the direction of the line of the section by consulting Table 1, in the Appendix.

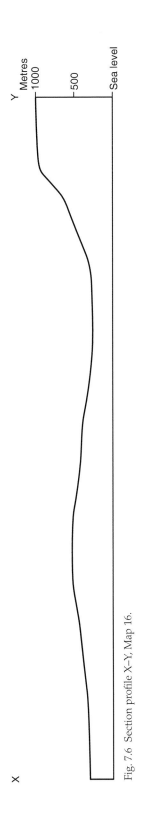

Fig. 7.6 Section profile X–Y, Map 16.

outcrops in the east. Erosional features have produced an inlier and an outlier.

Figure 7.2 is a simple map which shows beds outcropping in an area of simple topography: a north–south river valley with a high ridge to the west is confirmed by a number of spot-heights. Such information on the geology as is available shows an almost uniform dip of 11° to the south. The beds dip downstream (and are steeper than the slope of the valley floor), therefore their intersection with the topographic surface gives rise to V-shaped outcrops in the valley, that 'V' downstream. Naturally, the outcrops bend the other way over the ridge. Compare the outcrop pattern here with those on Maps 4 and 10.

Figure 7.3 is a map of a north–south valley sloping southwards and flanked either side by hills. Where beds dip upstream the outcrops 'V' upstream. Where beds dip downstream the beds 'V' downstream. This gives rise to lozenge- or 'diamond'-shaped outcrops, characteristic of folded strata where the fold axes tend to be more or less at right angles to the drainage. See, for example, Maps 12 and 21.

8

Faults

Chapter 7 showed how geological strata subjected to stress might be bent into different types of fold. Strata may also respond to stress (compression, extension or shearing) by fracturing.

Faults are fractures that displace the rocks. The strata on one side of a fault may be vertically displaced tens, or even hundreds, of metres relative to the strata on the other side. In another type of fault the rocks may have been displaced horizontally for a distance of many kilometres. While in nature a fault may consist of a plane surface along which slipping has taken place, it may also be represented by a zone of brecciated rock (i.e. rock composed of angular fragments). For the purposes of mapping problems it can be treated as a plane surface, usually making an angle with the vertical (see Plate 14).

All structural measurements are made with reference to the horizontal, including the dip of a fault plane. This is a measure of its slope (cf. the dip of bedding planes, etc.). The term 'hade', formerly used, is the angle between the fault plane and the vertical (and is therefore the complement of the dip). It causes some confusion but will be widely encountered in books and on maps and is, for this reason, shown in Figs. 8.1 and 8.2. It will, no doubt, gradually drop out of usage.

The most common displacement of the strata on either side of a fault is in a vertical sense. The vertical displacement of any bedding plane is called the throw of the fault. Other directions of displacement are dealt with later in this chapter.

Normal and reversed faults

If the fault plane is vertical or dips towards the downthrow side of a fault, it is called a normal fault

Plate 14 Fault, Kilve, Somerset. These Lower Jurassic (Liassic) limestones and shales are normally faulted with the right (northern) side downthrown. The limestone beds suggest movement in the opposite direction (upthrow on the right), but they do not correspond.

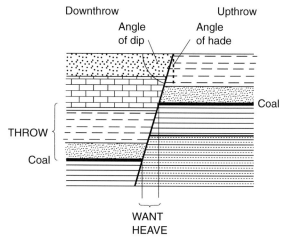

Fig. 8.1 Section through strata displaced by a normal fault (after erosion has produced a near-level ground surface).

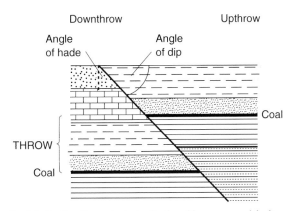

Downthrow   Upthrow

Angle   Angle
of hade   of dip

Coal

THROW

Coal

Fig. 8.2 Section through strata displaced by a reversed fault.

Plate 16 Small reversed fault with associated drag folds (above hammer head) in Oligocene limestones. Burranco Salado, near Altea, Spain.

(Fig. 8.1 and Plate 15). If the fault plane dips in the opposite direction to the downthrow (i.e. towards the upthrow side) it is a reversed fault (Fig. 8.2 and Plate 16).

In nature, the dip of a reversed fault is generally lower than that of a normal fault. It may be less than 45°. In an area of strong relief the outcrop of a reversed fault may be sinuous. The outcrop of a normal fault (commonly with a dip in the 85–75° range, but as low as 50° in some examples) will usually be much straighter – unless the fault plane itself is curved (see Map 19). Of course, where the area is of low relief, the outcrop of a fault plane, normal or reversed, will be virtually straight.

On some geological maps, such as those produced in Canada and the United States, the direction of dip of fault planes is shown. The angle of dip of the fault plane may be given in the map description. British Geological Survey maps show the direction of the throw of faults by means of a tick on the downthrow side of the fault outcrop.

Any sloping plane, including a fault plane, can be defined by its structure contours. It is possible on both Maps 17 and 20 to construct contours for the fault plane. The method is exactly the

Plate 15 Normal fault with downthrow to the left. Gibson Member, Upper Cretaceous Coals, McKinley Mine, North of Gallup, New Mexico, USA.

same as for constructing structure contours on bedding planes, described in Chapter 3. From these structure contours the direction of dip of the fault plane can be deduced and then, by reference to the direction of downthrow, it can be deduced whether the fault is normal or reversed.

The effects of faulting on outcrops

Consider the effects of faulting on the strata: those on one side of a fault are uplifted, relatively, many metres. Since this uplift is not as a rule a rapid process and the strata will be eroded away continuously, a fault may not make a topographic feature, although temporarily a fault scarp may be present (Fig. 8.3), especially after sudden uplift resulting from an earthquake.

Some faults which bring resistant rocks on the one side into juxtaposition with easily eroded rocks on the other side may be recognised by the presence of a fault line scarp (cf. the fault scarp resulting from the actual movement).

The strata that have been elevated on the upthrow side of a fault naturally tend to be eroded more rapidly than those on the downthrow side. This results in higher (younger) beds in the stratigraphical sequence being removed from the upthrow side of the fault while they are preserved on the downthrow side. It follows that we can usually determine the direction of downthrow of a fault, whether normal or reversed. Following the line of a fault across a map, there will be points where a younger bed on one side of the fault is juxtaposed against an older bed on the other side of the fault. The younger bed will be on the downthrow side of the fault.

A fault dislocates and displaces the strata. The effect of this, in combination with erosion, is to cause discontinuity or displacement in the outcrops of the strata.

Classification of faults

Faults may be categorised in two ways.

1. Faults may be classified according to the direction of displacement of the blocks of strata on either side of the fault plane. So far, we have considered normal and reversed faults with a vertical displacement called throw. Movement in these faults was in the direction of dip of the fault plane. They are called dip–slip faults because the movement – or displacement – was parallel to the direction of dip of the fault plane.

   On p. 69 faults with lateral displacement, wrench faults, are described. Here, displacement is parallel to the strike of the fault plane and they can be described as strike–slip faults.

   In nature, in some faults the displacement is neither dip–slip nor strike–slip but oblique. Naturally, the displacement will have a vertical component (throw) and a horizontal component (lateral displacement). Such a fault may be called an oblique-slip fault (Fig. 8.4).

2. A different classification of faults is dependent upon their geographical pattern, especially in relation to the dip directions of the strata cut by the faults. Where the faulting is parallel or nearly so to the direction of dip of strata, the faults are called dip faults. Where the faulting is more or less at right angles to the direction of dip of strata, i.e. approximately parallel to their strike, the

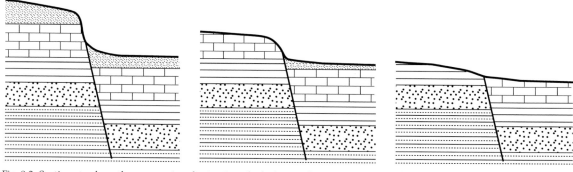

Fig. 8.3 Sections to show the progressive elimination of a fault scarp by erosion.

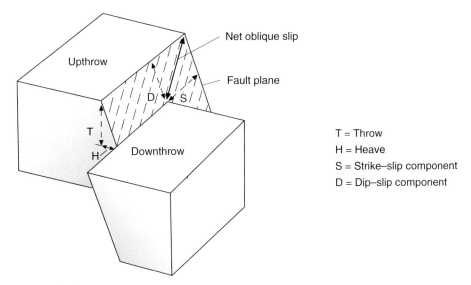

Fig. 8.4 An oblique-slip fault.

T = Throw
H = Heave
S = Strike–slip component
D = Dip–slip component

faults are called strike faults. Examples of both are found on Problem Map 19. Faults that are in neither the dip direction nor the strike direction may be called oblique faults.

It is particularly important not to confuse the two schemes of classification discussed above: the terms are unfortunately very similar. Test yourself: what is a dip–slip strike fault? (See block model 7 in the Appendix.) Figures 8.5 and 8.6 show examples of dip–slip faults; the former shows two cases of strike faults, the latter shows a dip fault. Figure 8.8 is an example of a strike–slip fault. It is, however, also a dip fault.

To return to normal dip–slip faults, sequences of outcrops encountered on a traverse may be partly repeated or may be partly suppressed.

Where the fault plane is parallel to the strike of the beds we see either repetition of outcrops (Fig. 8.5(a)) where the succession of beds at the surface reads A, B, C, A, B, C or the suppression of outcrops (Fig. 8.5(b)) where the succession of beds at the surface reads A, B, C, E, F, G.

Where the fault plane is parallel to the dip direction of the strata (a dip fault), i.e. at right angles to the strike, a lateral shift of the outcrop occurs. This must not be confused with lateral movement of the strata (see p. 69): the transposition of the

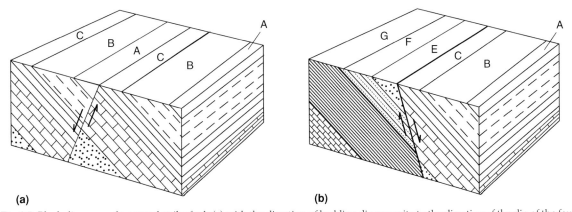

(a)   **(b)**

Fig. 8.5 Block diagrams of a normal strike fault (a) with the direction of bedding dip opposite to the direction of the dip of the fault plane, causing repetition of part of the succession of outcrops and (b) with the dips of bedding and the fault plane in the same direction, causing a suppression of part of the succession of outcrops.

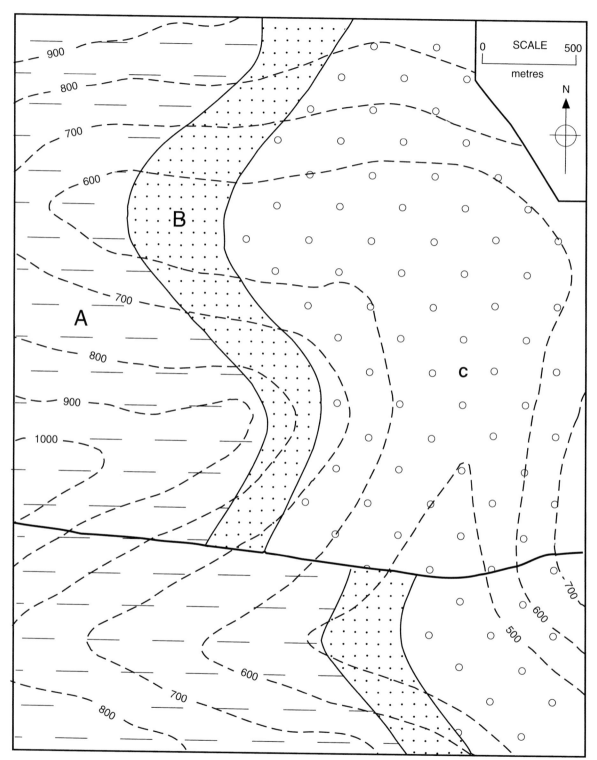

Map 17 Draw structure contours for the upper and lower surfaces of the sandstone (stippled). What is the amount of the throw of the fault? Draw structure contours on the fault plane. Is it a normal or a reversed fault? What is the thickness of the sandstone?

outcrops is due to vertical displacement of the beds followed or accompanied by erosion which, because the strata are inclined, causes the outcrops on the upthrow side to be shifted in the direction of dip (Fig. 8.6).

Calculation of the throw of a fault

Note on Map 17. Construct the structure contours on the upper surface of the sandstone bed in the northern part of Map 17. They run north–south and are spaced at 12.5 mm intervals. Follow the same procedure for the upper surface of the sandstone south of the fault plane. The 500 m structure contour drawn on the south side of the fault, if produced beyond the fault, is seen to be coincident with the position of the 1000 m structure contour on the north side of the fault. The stratum on the south side is, therefore, 500 m lower relatively. The fault has a downthrow to the south of 500 m.

Determine the throw of the fault using the structure contours on the base of the sandstone – on each side of the fault – and check that you obtain the same value.

Shade the areas on the map where a borehole would penetrate the full thickness of the sandstone. Also shade the areas where a borehole would penetrate only a partial thickness of the sandstone. You will find a zone where the borehole would not encounter the sandstone at all due to the heave (or want) of the fault. Note that this zone will be defined by constructing structure contours for the sloping fault plane as well as for the top and bottom of the sandstone. Intersections of these two sets of lines, where they are of the same height, will define where the sandstone is cut off by the fault plane. Joining these points of intersection will define the area(s) of absence, or partial absence, of the sandstone.

Faults and economic calculations

It can be seen that since the fault plane dips, the intersections with a bedding plane on each side of the fault do not coincide (in plan view). Consider an economically important bed, for example of coal or ironstone. In the case of a normal fault there is a zone where a borehole would not penetrate this bed at all due to the effect of heave (see Fig. 8.1). This is important when calculating economic reserves, for example in a highly faulted coalfield. The estimate of reserves could be as much as 15 per cent too great unless allowance had been made for fault heave. In the case of a reversed fault a zone exists where a borehole would penetrate the same bed twice. Fig. 8.7 shows, in section, how a borehole after penetrating a coal seam would penetrate the fault plane and, beneath it, the same seam. In calculating reserves it is vital to recognise that it is the same seam that the borehole has encountered or reserve calculations would be wrong by a factor of approximately 2.

Map 19 is a slightly simplified version of the western part of the BGS map of Chester, Sheet 109 in the 1:50,000 series, Solid edition. It is about half scale here. The geology of the area is relatively simple, with strata dipping generally towards the east with dips in the range of 5–17°. The topography is almost flat (and drift cover

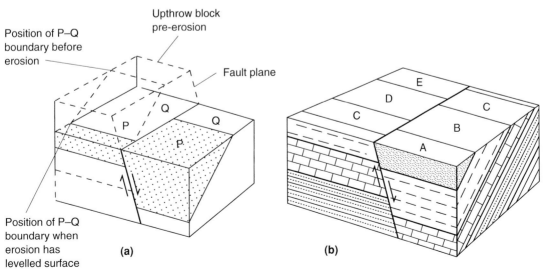

Fig. 8.6 (a) Dip fault showing how vertical throw gives rise to lateral shift of outcrop as a result of erosion. The outcrops shift progressively down dip as erosion lowers the ground surface. (b) Block diagram of a normal dip fault. Note the lateral shift of outcrops although the actual displacement is vertical.

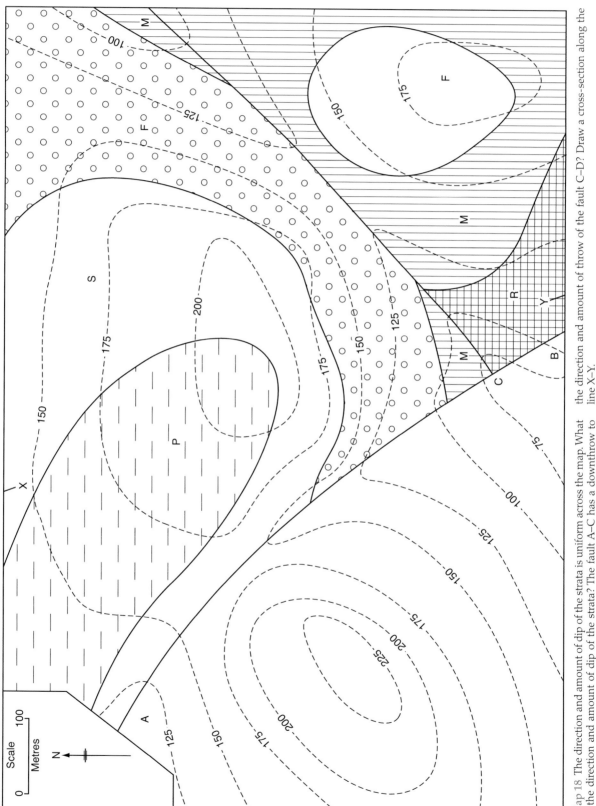

Map 18 The direction and amount of dip of the strata is uniform across the map. What is the direction and amount of dip of the strata? The fault A–C has a downthrow to the north-east of 50 m. Complete the outcrops to the south-west of this fault. What is the direction and amount of throw of the fault C–D? Draw a cross-section along the line X–Y.

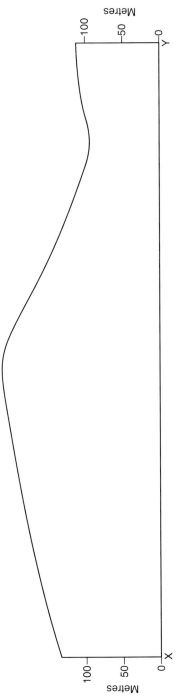

Fig. 8.7 Profile for Map 18.

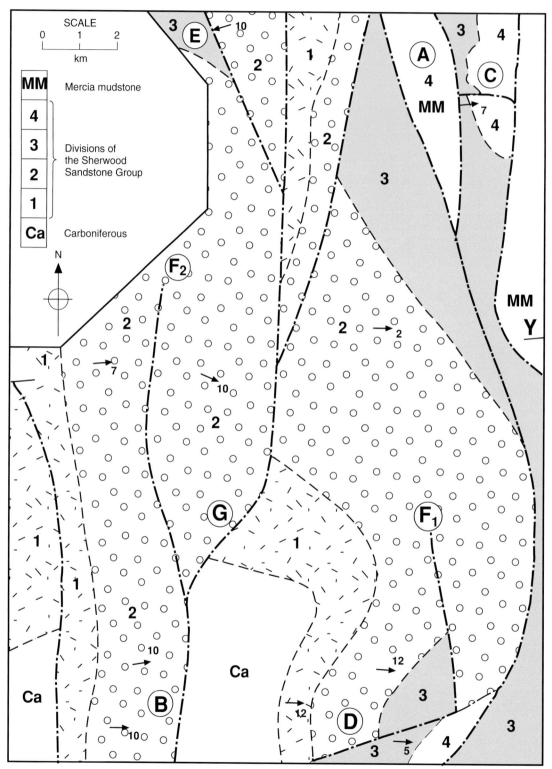

Map 19 On the map indicate the downthrow direction of all faults where it has not been shown. Indicate examples of (a) a graben and (b) step faulting. Draw a section along the nearly east–west line X–Y to illustrate the geology. All faults are nor-mal faults, assume that they dip (hade) towards the down-throw. Reproduced by permission of the Director General, British Geological Survey: NERC copyright reserved.

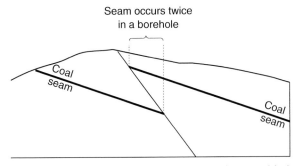

Fig. 8.8 Section to show a 'low angle' (high hade) reversed fault and its importance in mining problems.

is extensive so that much of the map has been compiled using borehole data). Strata are displaced by a considerable number of faults and they display many important characteristics of faulting.

Note that faults may die out laterally; examples can be seen at F_1 and F_2. Of course, all faults die out eventually, unless cut off by another fault, as at C, though major faults may extend for tens or even hundreds of kilometres. Note also that a fault may curve, for example at G. This is not mere curvature of the outcrop of the fault plane due to the effects of topography on a sloping fault plane (see Map 20).

Most of the faults on Map 19 run approximately north–south, roughly parallel to the general strike of the strata: they can be called strike faults. The displacement is in the direction of dip of the fault planes so they can also be called dip–slip faults. Two faults are approximately parallel to the direction of dip of the strata, seen at C and D. These are, therefore, dip faults. They are also probably dip–slip faults. The Chester sheet indicates the direction of downthrow of each fault, as do most

published maps, but this has been omitted from most of the faults on Map 19.

Wrench or tear faults

In the case of these faults the strata on either side of the fault plane have been moved laterally relative to each other, i.e. movement has been a horizontal displacement parallel to the fault plane. In the case of simply dipping strata the outcrops are shifted laterally (Fig. 8.8) so that the effect, *on the outcrops*, is similar to that of a normal dip fault (see Fig. 8.6) – and in this case it is usually impossible to demonstrate strike–slip from the map alone (apparent slip only can be found).

In some geological contexts the terminology based on the direction of movement relative to the dip and strike of the fault plane is preferable. Faults in which movement has been in the strike direction of the fault plane (= strike–slip faults) include wrench faults.

Pre- and post-unconformity faulting

After the deposition of the older set of strata, earth movements causing uplift may also give rise to faulting of the strata. The unconformable series (the younger set of beds), not being laid down until a later period, is unaffected by this faulting. Earth movements subsequent to the deposition of the unconformable beds would, if they caused faulting,

Notes on Map 19. The dip of the strata at E is anomalous, the result of a phenomenon called fault-drag.

The effect of faults which are parallel to the dip of the strata is to laterally shift outcrops (see p. 65 and Map 17). The extent of this shift depends on two factors: the amount of throw of the fault and the angle of dip of the strata. At C, on Map 19, the geological boundary is displaced very little, but at D the displacement is considerable (the shift is almost equal to the width of outcrop of Bed 4). At C and D the dips of the strata are similar so we can conclude that the fault at D has a much larger throw than the fault at C.

It may be assumed that all the faulting here is normal. Since we have no means of calculating the dip of the fault planes, in your section give faults a conventional dip towards the downthrow of 80–85°.

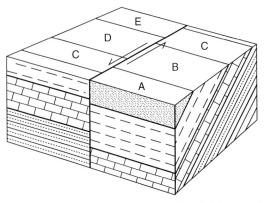

Fig. 8.9 Block diagram of a wrench (= tear) fault. Note that the effect on the outcrops is similar to that of a normal dip fault (cf. Fig. 8.6(b)).

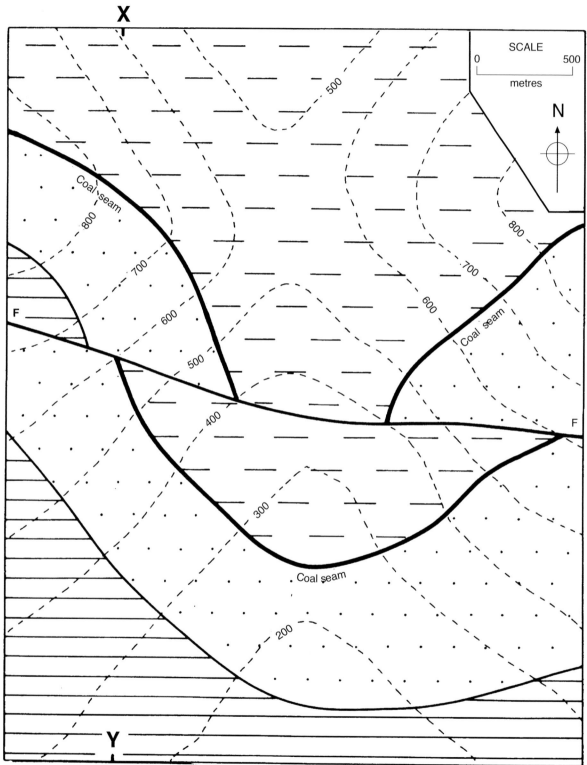

Map 20 The line F–F is the outcrop of a fault plane. The other thick line on the map is the outcrop of a coal seam. Shade areas where coal could be penetrated by a borehole (where it has not been removed by erosion). Indicate areas in which any borehole would penetrate the seam twice. What type of fault is this? Draw a section along the line X–Y. Draw the 100 m overburden isopachyte, i.e. a line joining all points where the coal is overlaid by 100 m of strata. (See also pp. 69.)

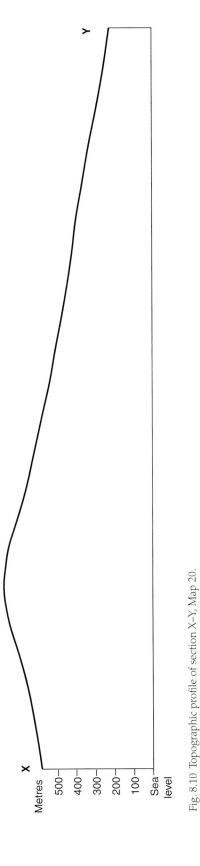

Fig. 8.10 Topographic profile of section X–Y, Map 20.

produce faults that affect both sets of strata. Clearly, it is possible to determine the relative age of a fault from inspection of the geological map which will show whether the fault displaces only the older (pre-unconformity) strata or whether it displaces both sets of strata. A fault is later in age than the youngest beds it cuts.

A fault may also be dated relative to igneous intrusions, a topic dealt with in the final chapter.

Structural inliers and outliers

The increased complexity of outcrop patterns due to unconformity and faulting greatly increases the potential for the formation of outliers and inliers (these terms have been defined in Chapter 3). Indicate on Map 21 inliers and outliers that owe their existence to such structural features and subsequent erosional isolation.

Posthumous faulting

Further movement may take place along an existing fault plane. So the displacement of the strata is attributable to two or more geological periods. It follows that an older series of strata may be displaced by an early movement of the fault which did not affect newer rocks since they were laid down subsequently. The renewed movement along the fault will displace both strata so the older strata will be displaced by a greater amount since they have been displaced twice (the throws are added together).

Isopachytes

Isopachytes (*iso* = equal; *pakhus* = thick) are lines of equal thickness. The simplest use of isopachytes is to show the thickness of cover material overlying a bed of economic importance, such as a coal seam or ironstone. The overlying material – whatever its

Notes on Map 20. The zone in which a borehole penetrates the seam twice is defined by the lines of intersection of the fault plane and the coal seam (Fig. 8.7). The surfaces intersect where they are at the same height (where the coal seam structure contours and the fault plane structure contours of the same height coincide).

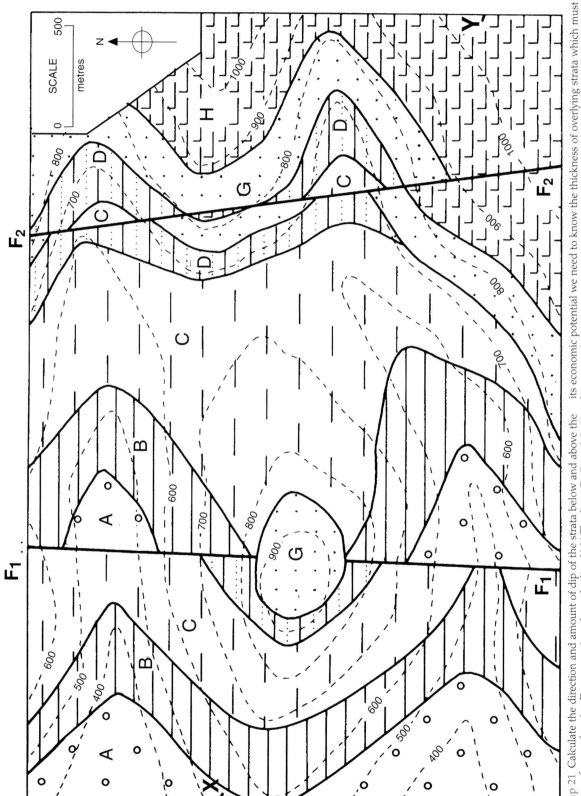

Map 21 Calculate the direction and amount of dip of the strata below and above the plane of unconformity. Draw a section along the line X–Y. The lines F_1–F_1 and F_2–F_2 are the outcrops of two fault planes. Which fault occurred earlier in geological time? Assume that Bed B is a commercially important bed of ironstone. In order to estimate its economic potential we need to know the thickness of overlying strata which must be removed in order to mine the ironstone by opencast methods (strip mine). Draw on the map the 100 m and 200 m isopachytes for this cover (overburden).

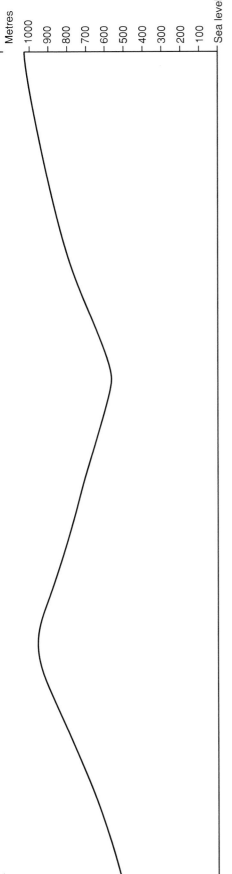

Topographic profile of section X–Y, Map 21.

composition: strata, soil or subsoil – is called the overburden. Its thickness can be determined where the height of the top of an economic bed (ironstone, Map 39) is known from its structure contours, and the height of the ground at the same point is known from the topographic contours. Wherever structure contours and topographic contours intersect on the map we can obtain a figure for the thickness of overburden (by subtracting the height of the top of the ironstone from the height of the ground). Joining up the points of equal thickness gives an isopachyte. Where ironstone and ground are at the same height the thickness of overburden is nil and the bed must outcrop. (Its outcrop would be the 0 m isopachyte.) Bed (or stratum) isopachytes, concerned with beds of varying thickness, are dealt with in Chapter 10.

Exercises on published geological survey maps

Chesterfield: 1″ Geological Survey map (Sheet 112) Solid & Drift edition, 1963. Find the major unconformity on this map. Locate some faults that are older than the Permian beds and some that are younger. Draw a section along a line across the map ensuring that it passes through the Ashover anticline in the south-west.

Leeds: 1:50,000 (Map 70) Solid & Drift edition, 1979. Draw a section along the 'Line of Section' engraved on the map to show the geological structures.

Stockport: 1:50,000 (Sheet 98) Solid edition, 1979. Note the faulting *trend*. Faulting is generally north–south or slightly west of north to east of south. (A few faults trend NE–SW.) The Red Rock Fault (and its branches) drops the Triassic strata down to the west to form the Cheshire Plain, downfaulted against the gently folded Westphalian (Carboniferous) Coal Measures. Look for faults responsible for repetition of outcrops. There is one example of a horst, a block upfaulted relative to the strata on either side (the town of Cheadle sits partly on it). Note that some faults die out laterally, some faults are cut off by another (which must be of slightly later date).

This is an area covered largely by glacial drift and the existence and location of many of the faults have been proved by boreholes for salt or coal.

9

Map solution without structure contours 2

The methods employed in the solution of maps without recourse to attempting to draw structure contours were dealt with in Chapter 7.

This chapter includes more difficult maps, which incorporate faults dealt with in Chapter 8, as well as all structures discussed so far. Some maps also include igneous features dealt with in Chapter 11.

In addition, you will find it useful to refer to maps without structure contours, such as Maps 35, 36 and 37 in Chapter 11.

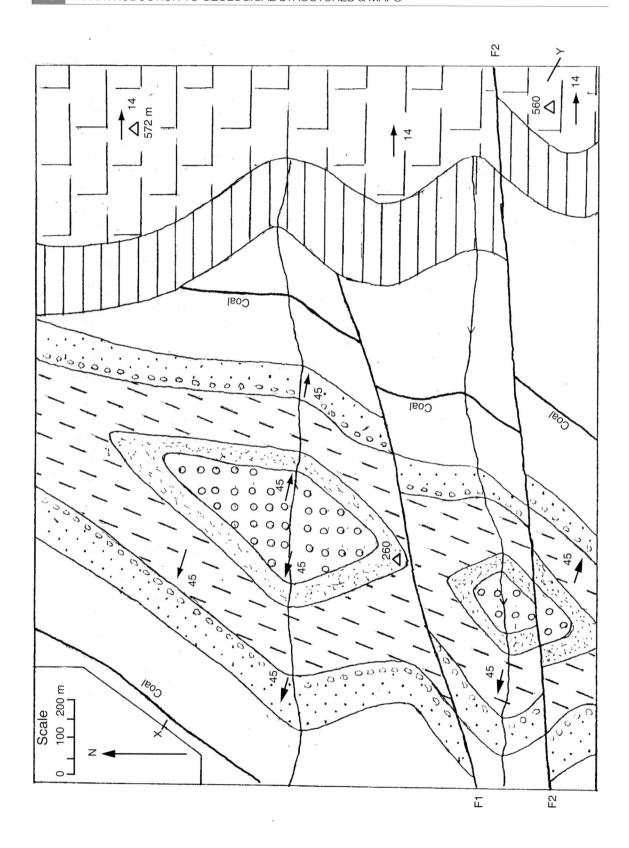

Map 22 The map shows an area of folded and faulted Carboniferous age Coal Measures. Shading indicates shales, sandstone with a basal conglomerate, grit and a thick coal seam. To the east the Coal Measures are overlain by basal marls and limestone of Permian age. Insert a fold axis. There are two faults. Both have a throw of 100 m, but what is their relative age?

Draw a section along the line X–Y on the profile provided to show the geology. Assume that both faults are normal and give them a dip of 85° in the direction of their downthrow. To the south–east of fault F$_2$ you need insert on the section only the coal seam. The horizontal and vertical scales are the same so there is no vertical exaggeration affecting the dips. However, if the direction of dip of the beds does not coincide with the direction of the section line then we will see in the section an apparent dip. The angle between the direction of dip in the Coal Measures and the line of section is only 15°, so the dip in the line of section is only a degree less than true dip. (See graph, Figure XX [orig 51] in the Appendix.)

The dip of the Permian makes an angle of 30° with the line of section, which means that the apparent dip seen in the line of section will be reduced. Since the Permian has a low dip of only 14°, it will be reduced by only a couple of degrees. Again consult graph, Figure XX [orig 51].

Reminder: Tan 14° × Cos 30° (0.249 × 0.8660 = Tan^{-1} 0.216 (12°).

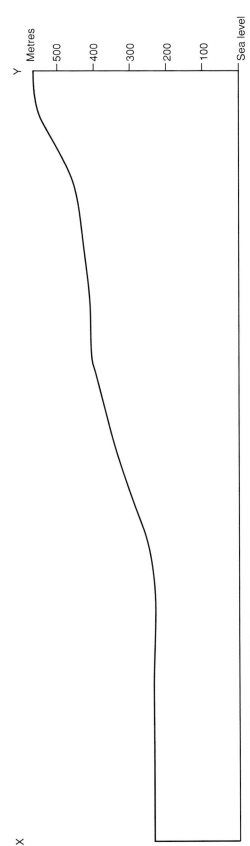

Fig. 9.1 Topographic profile of section X–Y, Map 22.

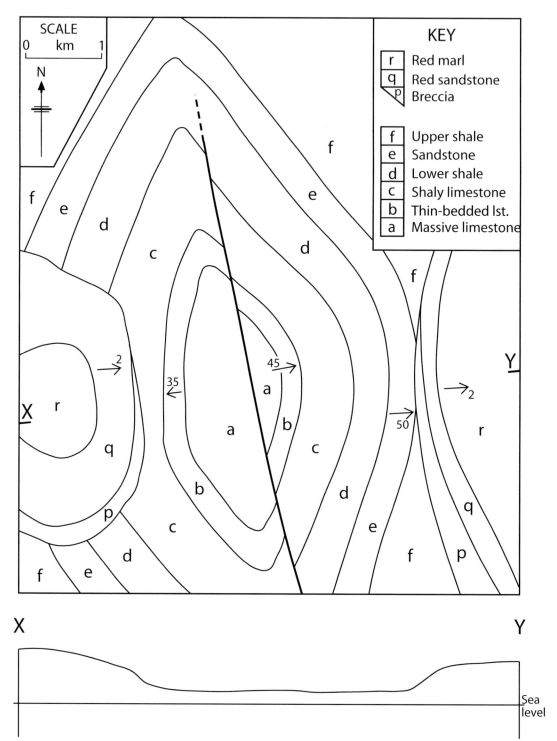

Map 23 Describe the geology of the area of the 1:50,000 scale map. Draw a section along the line X–Y on the profile provided. Show the structures to a depth of 250 m or more below sea level (which from a structural geology viewpoint is a quite arbitrary datum).

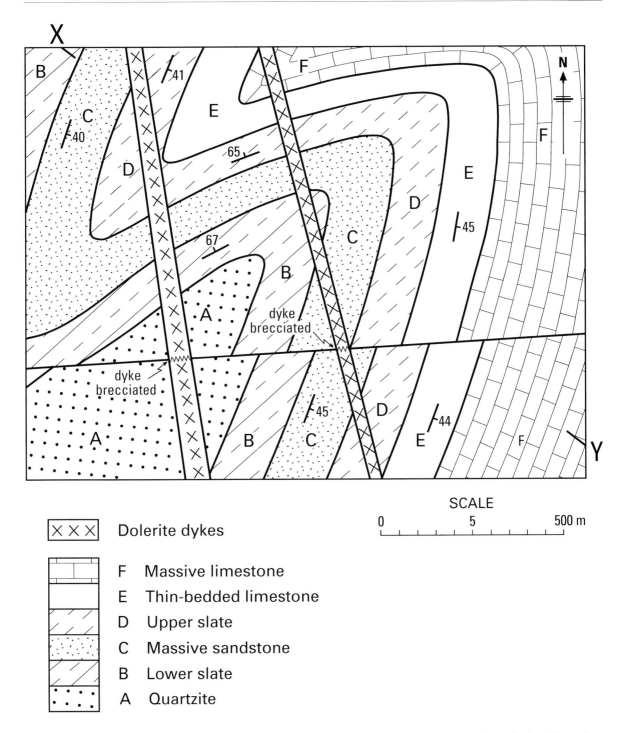

SCALE

0   5   500 m

☒☒☒	Dolerite dykes

	F	Massive limestone
	E	Thin-bedded limestone
	D	Upper slate
	C	Massive sandstone
	B	Lower slate
	A	Quartzite

Map 24 The map shows the geological exposures on a small area of wavecut platform. The surface is nearly horizontal and almost flat, exposure is very good (except for seaweed and fallen blocks from the adjacent cliffs, which have been omitted to simplify the map). Draw a section along the line X–Y to show the structures present, assuming a flat ground surface. Use the same vertical scale as the horizontal scale, i.e. no vertical exaggeration, so that angles of dip will be correctly represented.

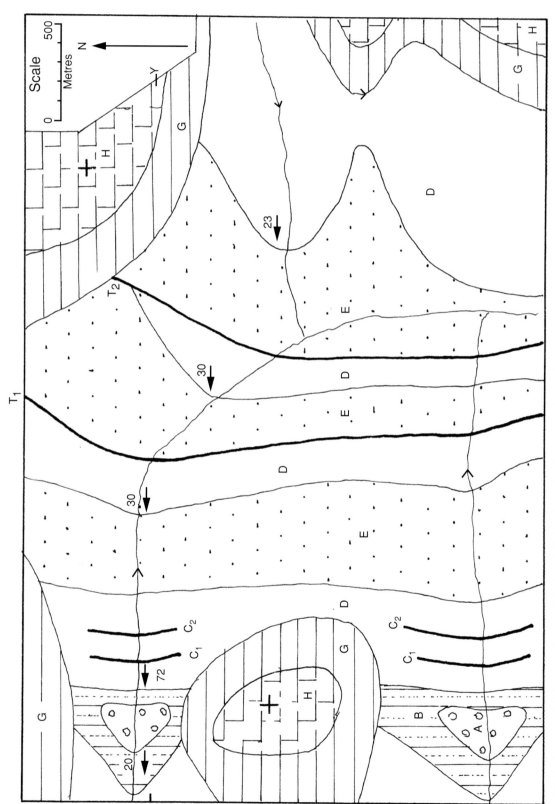

Map 25 In this area of slight to moderate relief, there is little relationship between geology and topography apart from the younger horizontal beds (G and H) forming the high ground. Bed G rests unconformably on an older series of strata that has been folded and thrust. Assume that there is field evidence showing that both the thrust planes are dipping westwards at 45°. C_1 and C_2 are coal seams found locally in shale D.

Draw a section along the line X–Y on the profile provided and comment on the structures.

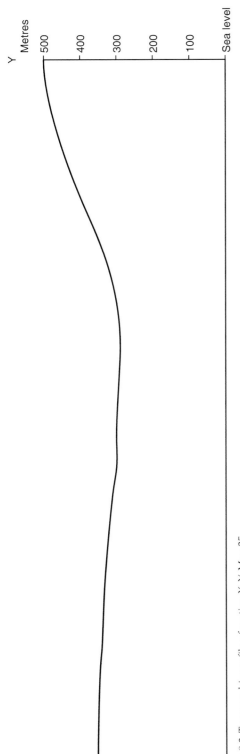

Fig. 9.2 Topographic profile of section X–Y, Map 25.

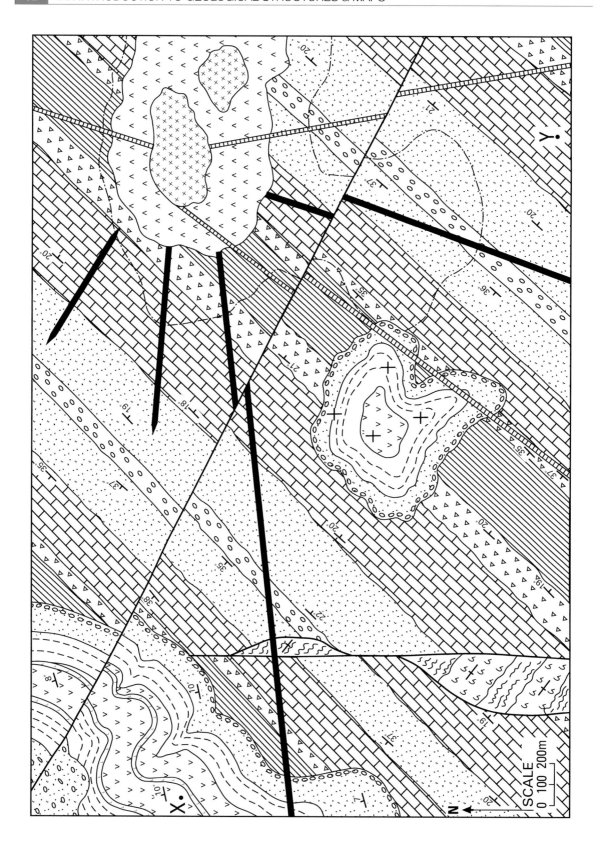

SCALE
0 100 200m

Map 26 There are two faults: one runs north–south and the other runs north-west to south-east. Determine, from the outcrop patterns, whether the displacement has been vertical or horizontal in each case. Are the folds symmetrical or asymmetrical? What is the evidence? Mark on the map the major unconformity and indicate inliers and an outlier. Can you decide in which direction the lower lava flowed? From the metamorphic aureole, which side of the plutonic intrusion has the steepest side? What surface evidence gives you a clue? Indicate on the map where a still buried pluton may lie.

Describe the geology of the area of the map. Draw a section along the line X–Y on the profile provided. The geological events that occurred in the area can be relatively dated and then arranged in chronological order, providing a summary of the geological history. For further guidance on this refer to 'Description of a Geological Map' on p. 136.

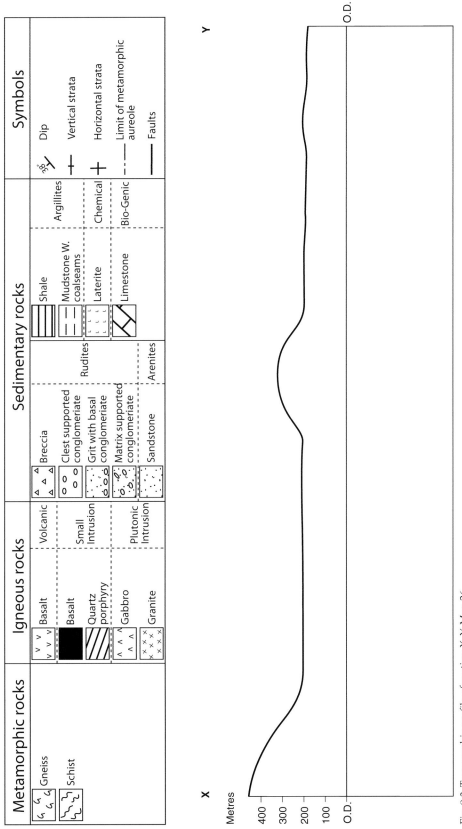

Fig. 9.3 Topographic profile of section X–Y, Map 26.

Note on Map 23. Note the asymmetry of the fold, which is a pericline. Note that the map includes an example of overlap as well as overstep. The throw of the fault decreases northwards and it dies out within the outcrop of Bed d. Of course, no faults go on for ever but faults of variable throw introduce problems usually reserved for advanced textbooks.

Note on Maps 24 and 26. There are some structural elements in Maps 24 and 26 that we have not yet dealt with. You may be able to deduce the structures present but, at least in the case of Map 26, you are strongly advised to read Chapter 11 on igneous rocks before attempting to answer the questions on this map.

10

More folds and faulted folds

Plunging folds

In the folds studied so far, the axis of the fold has been horizontal. The axis is the intersection of the axial plane with any bedding plane. (Make a synclinal fold by taking a piece of paper and folding it in two to make a simple V-shape. Drop a pencil into this and it will assume the position of the axis of the fold.) Such a fold, with the axis horizontal, is called a 'non-plunging' fold. The outcrops of the limbs tend to be parallel (but of course are affected by the configuration of the ground) since the structure contours drawn on one limb are parallel to those drawn on the other limb. (They have, as well, been parallel to the axial plane and, in cases where the axial plane was vertical, parallel to the axial trace (= outcrop of the axial plane).)

Where the fold axis is not horizontal, but is inclined, the fold is described as plunging. Some fold structures can be traced for many kilometres, others are of much more limited extent. Eventually a fold will die out by plunging (Fig. 10.1). Anticlinal folds that die out by plunging at both ends, giving rise to elliptical outcrop patterns, are called periclines and are a common occurrence.

Simple folds that have been subsequently tilted by further earth movements will also plunge. This is the simplest way of considering such structures, although they may originate in other ways. This type of fold is referred to in some earlier books as a 'pitching' fold, and not infrequently the terms 'pitch' and 'plunge' are used synonymously. The advanced student should consult FC Phillips' *The Use of Stereographical Projection in Structural Geology*, 3rd edn (Hodder Arnold, 1971), which deals fully with these terms. It is best to restrict the term 'pitch' to such linear features as striae, slickensides or lineations on a plane. They have both plunge (the angle they make with the horizontal) and pitch (the angle they make with the strike direction of the plane). Reject the use of the term pitch when referring to folds. Since problems entailing pitch can be solved (quite simply) by the use of stereographic projection, they are not included in this book.

The effects of erosion on plunging folds are seen in Figs. 10.2 and 10.3. The outcrops of the geological interfaces of the two limbs of a fold are not parallel. While the structure contours are parallel for the beds of each limb, those of the two limbs converge, meeting at the axial plane (Fig. 10.4).

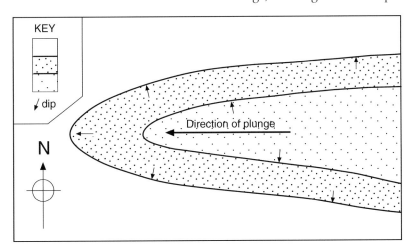

Fig. 10.1 Map showing the outcrops of anticlinally folded beds, plunging to the west.

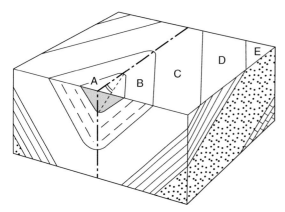

Fig. 10.2 Block diagram of a plunging syncline.

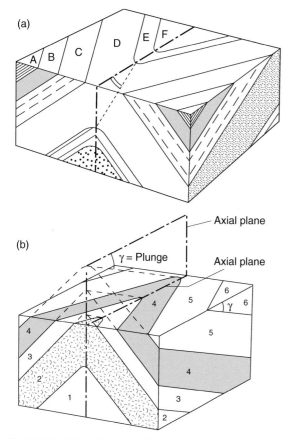

Fig. 10.3 (a) Block diagram of a plunging anticline with the plunge towards the observer. (b) Plunging anticline with the plunge away from the observer, eroded strata shown in broken lines. (See block model 4 in the Appendix.)

Calculation of the amount of plunge

Just as the inclination of the beds (dip) can be calculated from the spacing of the structure contours

measured in the direction of dip (see Chapter 3, 'Section drawing'), so can the plunge of the fold be calculated from the spacing of the structure contours measured in the direction of plunge, i.e. along the axis. The plunge of the fold shown in Fig. 10.4 is, expressed as a gradient, 1 in 3.5 (to the north), if the scale of the diagram is 1 cm = 200 m, since the structure contour spacing measured along the axis is 1.75 cm. (The dip of the bedding plane of each limb is 1 in 2: check that this is so.)

Where we have no structure contours enabling accurate calculation of the angle of plunge, it is possible to give some estimate of it from the width of outcrop. In Fig. 10.1 the outcrop of the coarsely stippled bed is about twice as wide in the direction of plunge as it is in the limbs of the fold. We can conclude that the amount of plunge is considerably less than the angle of dip of the limbs of the fold. The relationship of dip to width of outcrop was discussed in Chapter 3.

The effects of faulting on fold structures

We have seen in Chapter 8 that the effect of dip–slip faulting (normal and reversed faults), followed by erosion, produces a lateral shift of outcrop in the direction of the dip of the strata on the upthrow side. Observe that on Fig. 10.5 the vertical axial plane of a symmetrical fold is not displaced but the dipping planes show a shift of outcrop. Remember that the lower the angle of dip, the greater the shift of outcrop.

Since the shift of outcrop is down-dip as a result of erosion, it follows that on the upthrow side of a fault the outcrops of the limbs of an anticline are more widely separated (than on the downthrow side). Conversely, the outcrops of the limbs of a syncline are closer together on the upthrow side (than on the downthrow) (Fig. 10.5).

On an asymmetrical fold, there will be a more pronounced shift of the outcrops of the limb with the lower angle of dip and a less pronounced shift of the outcrops of the steeper dipping limb. In the 'limiting' case of a monoclinal fold with a vertical limb there will be no shift of outcrops of the vertical beds. The axial plane of an asymmetrical fold dips, therefore the axial trace will be displaced.

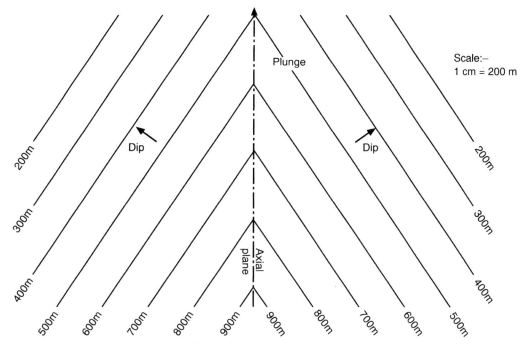

Fig. 10.4 Map of the structure contour pattern of a plunging fold.

Plate 17 Plunging open syncline, Kilve, Somerset. This wave-cut shore exposes a gentle fold with shallow dipping limbs. The harder limestone beds tilt inwards, indicating a syncline. The curved pattern of strata shows that the syncline plunges towards the bottom left. A fault runs across the centre of the picture, slightly displacing the nearest limb of the fold.

In the case of overfolds – where both limbs dip in the same direction (see Fig. 6.9) – naturally the outcrops of both limbs will be shifted in the same direction by a dip–slip fault. Of course, the outcrops of the shallow limb will be shifted more than those of the steeper dipping overturned limb.

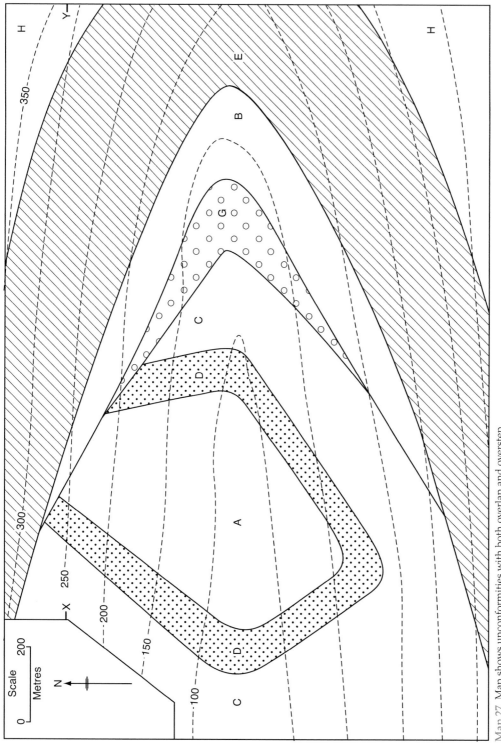

Map 27 Map shows unconformities with both overlap and overstep.

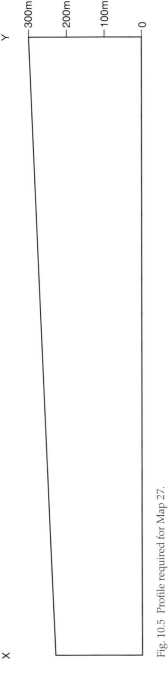

Fig. 10.5 Profile required for Map 27.

Questions on Map 27

1. Calculate the direction and amount of plunge shown by the Beds C, D and A.
2. Draw a cross-section on the profile X–Y provided.
3. Does Bed C overlap or overstep onto Bed D?
4. Does Bed B overlap or overstep onto Bed A?

Notes on Map 27. This map illustrates the concepts of both overlap and overstep related to unconformities (see p. 30).

By drawing strike lines on E/B contact and then a second set of strike lines on both the western and eastern outcrops of the D/A contact, the essential structures within the area will become clearer.

Displacement of folds by strike–slip (wrench) faults

A wrench fault causes a lateral (horizontal) dislocation or displacement of the strata, which may in some case, be of many kilometres. The effect of a wrench fault is to displace the outcrops of beds, always in the same direction – and its effect on the outcrops of simply dipping strata in certain circumstances is similar to that of a normal fault: it may not be possible to recognise from the evidence a map can provide whether a fault has a vertical or a lateral displacement (see p. 69). However, the effects of a wrench fault on folded beds are immediately distinguishable from the effects of a normal fault since it will displace the outcrops of

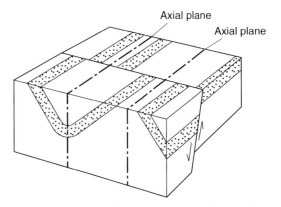

Fig. 10.6 Block diagram showing folded strata (one bed is stippled for clarity) displaced by a normal fault.

both limbs by an equal amount and in the same direction (whether the fold be symmetrical or asymmetrical) and, further, the outcrops of a bed occurring in the limbs of a fold will be the same distance apart on both sides of the fault plane (Fig. 10.6). The axial plane will, of course, be laterally displaced by the same amount as the outcrops of the beds occurring in the limbs of the fold, and this displacement occurs whether the axial plane is vertical (in a symmetrical fold) or inclined (in an asymmetrical fold). Compare Figs. 10.6 and 10.7.

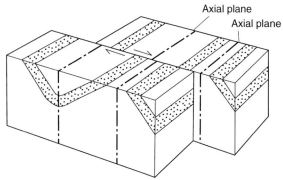

Fig. 10.7 Block diagram showing similar beds to Fig. 10.6 displaced by a wrench fault.

Calculation of strike–slip displacement

Where any vertical phenomenon is present, for example the axial plane of a symmetrical fold, vertical strata, igneous dyke, etc., the lateral displacement by the fault can be measured directly from the map. If the strata are dipping and the fault is a purely strike–slip fault with no vertical component, or throw (i.e. it is not an oblique-slip fault, see Fig. 8.4), then we can simply find the lateral shift by measuring the displacement of any chosen structure contour on a given geological boundary. Where strata are folded it is easy to measure the lateral displacement of fold axes or axial planes (Fig. 10.7).

Faults parallel to the limbs of a fold

So far we have considered faults that were perpendicular to the axial planes of the folds or, at least, cut across the axial plane. A fault which is parallel to the strike of the beds forming the limb of a fold is, of course, a strike fault. This will cause either repetition of the outcrops or suppression of the outcrops of part of the succession, in the manner discussed on p. 63.

Sub-surface structures

An interesting and important geological consideration is the deduction of the disposition of strata beneath an unconformity. In effect we are considering the 'outcrop' pattern on the plane of unconformity. If we could use a bulldozer to remove all strata above that plane we would see the earlier, pre-unconformity strata outcropping.

This problem was touched on briefly in Map 10 where unconformity was introduced. If not already completed, return to Map 10 and insert the sub-unconformity outcrop of the coal seam. (Remember that topographic contours are now irrelevant; the surface we are considering is the plane of unconformity which is defined by the structure contours drawn on the base of Bed Y. Since both sets of structure contours are in each case straight, parallel and equally spaced – representing constant dips – their intersections giving the sub-Y outcrop of the coal seam should lie on a straight line.)

Posthumous folding

After the strata in an area have been laid down they may be uplifted, folded and eroded, as we have seen in Chapter 9. Further subsidence may cause the deposition of strata lying unconformably upon existing beds. Later still the processes of uplift and folding may recur, when this second period of folding is said to be posthumous – if the fold axis in the younger beds is parallel to, and approximately coincident with, the similar fold axis in the older beds. The trend of the two sets of folds may be parallel and the fold axes may coincide, as in Fig. 10.8, or they may be parallel but not coincident. An excellent example of this is demonstrated by the Upper and Lower Carboniferous age beds in the Forest of Dean area included on the BGS 1:50,000 map of Monmouth (Sheet 233).

Since the folding is of two ages, the trends of the two sets of folds may be in quite different directions and it is then called superposed folding or

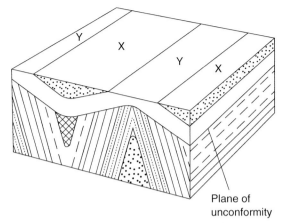

Fig. 10.8 Block diagram showing the effects of posthumous folding.

cross-folding and is not included in posthumous folding; however, there is usually a tendency for the earlier folding to exercise a 'control' over the later folding so that, more commonly, the trends are parallel or sub-parallel.

The strata of Pembrokeshire (see the Haverfordwest and Pembroke 1″ Geological Survey maps (Sheets 227, 228)) were affected by Caledonian folding and later by Variscan (Hercynian) folding. In this area the trend of the folding of both periods is nearly east–west. By contrast, the rocks of the Lake District were gently folded with an east–west strike in Ordovician times and later by the Caledonian orogeny (phase in the mountain-folding process) with a north-east–south-west strike, the later folding being of such intensity that it seems to have been independent of the control of the earlier folds.

Where the folds are of different trend it is possible to study the effects of one age of folding by taking a section parallel to the strike of the other folds. This was very neatly shown in the block diagram of a part of the southern Lake District in earlier editions of the 'Northern England' British Regional Geology.

Polyphase folding

Strata that have already been folded may at a later time be refolded. The stress causing the refolding may be due to a later phase of the same orogeny or even due to a later orogeny. The stress of the later phase of folding may be quite unrelated to the stress direction of the earlier folding (the 'first folds'). As a result of this the trends of the two ages of folding may be quite different.

Complex outcrop patterns are produced by the interference of two (or more) phases of folding. Simple examples are illustrated (Fig. 10.9). In some such simple cases the early folds can be seen to have been refolded since the axial planes of the first folds are themselves folded.

Bed isopachytes

All problems so far have dealt with beds of uniform thickness. However, traced laterally over some distance, strata may be seen to vary in thickness, a function of their mode of deposition. Such variations tend to be gradual and reasonably uniform within the area of a geological map sheet. Variations in the thickness of a bed are usually deduced

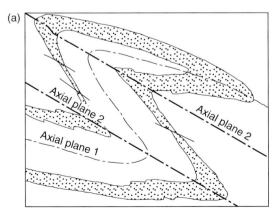

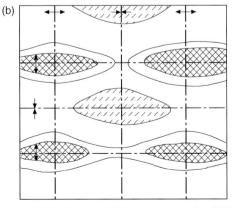

Fig. 10.9 Outcrops of refolded folds on flat ground. (a) A refolded isoclinal fold, (b) second fold axes crossing first fold axes at right angles.

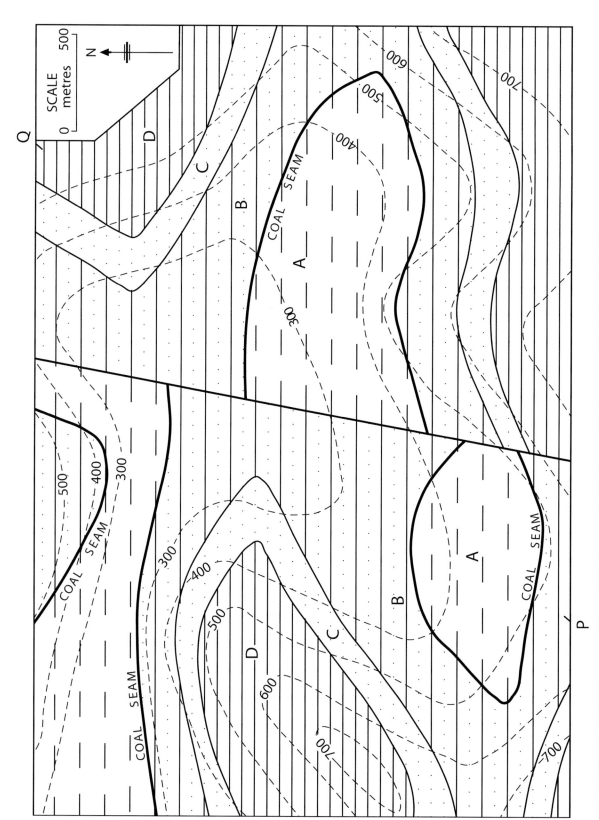

Map 28 What kind of fault occurs on this map? Has it any vertical displacement?   the map anticlinal and synclinal axial traces.
Draw a section across the map along the line P–Q intersecting the fault. Indicate on

from borehole data but may be discovered by measuring sections at geological outcrops and are occasionally revealed by variations in width of outcrop on a map.

The way in which a bed varies laterally in thickness can best be shown by constructing a series of bed isopachytes, lines joining points where the bed is known to be of the same thickness. To obtain the maximum number of control points at which the thickness of a bed can be determined, it is usually necessary to construct two sets of structure contours: those for the top of the bed and those for its base. Their intersections give the thickness of the bed. Due to the nature of the sedimentary phenomena, which produce beds of varying thickness, the isopachytes will tend to be reasonably straight (or gently curving) and approximately evenly spaced.

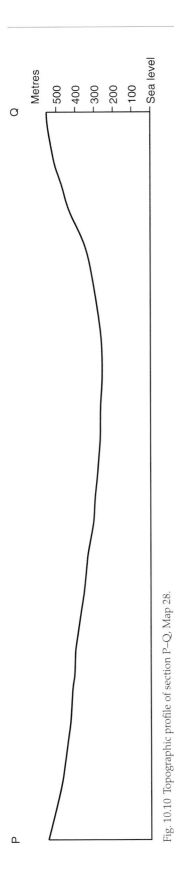

Fig. 10.10 Topographic profile of section P–Q, Map 28.

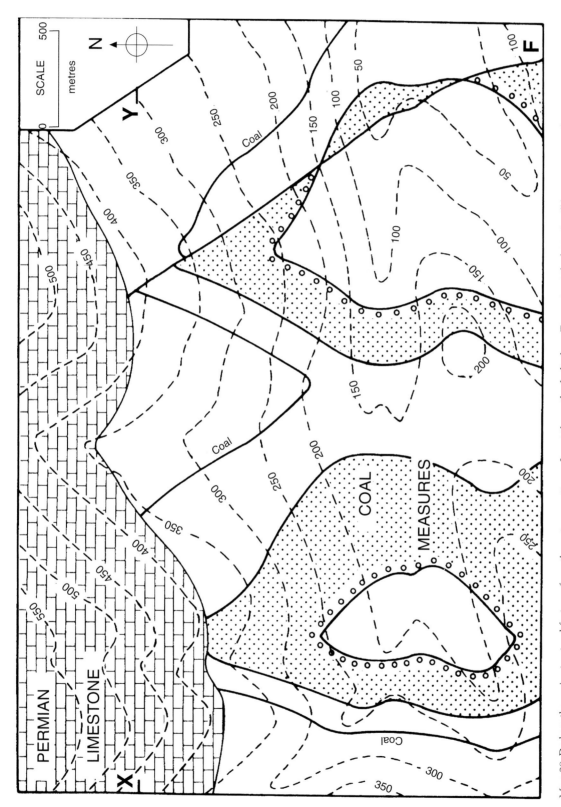

Map 29 Deduce the main structural features from the outcrop patterns. Insert the fold axial traces on the map. Draw a geological section along the line X–Y. Insert on the map the sub-Permian Limestone outcrops of the coal seam, the sandstone and the fault plane. To construct the latter it will be necessary to draw structure contours for the fault plane in order to deduce its dip.

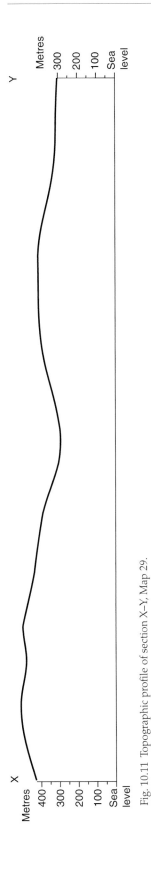

Fig. 10.11 Topographic profile of section X–Y, Map 29.

1. **Monmouth: 1 inch to the mile (Sheet 233) Solid & Drift edition, 1974**. Note the large fold structure in the eastern quarter of the map affecting Coal Measures. This is clearly a synclinal structure with the youngest beds at the centre. It outcrops close to the north and to the south, indicating that it is an elongate basin, essentially.

   Note that the Lower Carboniferous beds are also folded synclinally but the fold axis (just to the west of Lydney) does not precisely coincide with the axis in the Upper Carboniferous. Here we have an example of posthumous folding (despite the fact that the axes are not quite coincident): during the period of time represented by the unconformity, the Lower Carboniferous beds were folded. After the Upper Carboniferous beds were deposited, folding about a north–south axis resumed. Note that the Coal Measures overstep the beds of the Lower Carboniferous age, also confirming the unconformable relationship.

2. **Haverfordwest: 1:50,000 (Map 228) Solid & Drift edition, 1976**. Draw a section along the north–south grid line 197. Note the difference in amplitude of the folding in the Lower and Upper Palaeozoic rocks. Tabulate the geological events in chronological order.

Plate 18 Refolded folds, Loch Monar, Scotland. These Pre-Cambrian schists have been folded with horizontal axes. A second phase of folding has superimposed closer spaced folds, with vertical axes, onto the earlier larger folds. The picture is about 1.5 m across.

Problem 2

Four boreholes were drilled at 500 m intervals along a straight line. The vertical sections reveal in each case a succession of sedimentary rocks. Assuming that similarity of lithology and thickness of a bed indicate a probable correlation, show on the diagram the geological structures which may be present to explain the relationship of the strata in one borehole to the strata in the adjacent boreholes.

Note on Problem 1. The higher strata in all four boreholes are the same in sequence and in having no dip – they are horizontal. Below the limestone it is clear that the beds are folded, proving the presence of an unconformity. In Boreholes A, B and C the relationship can be simply shown by sketching in the appropriate fold structure. In Borehole D the dip of the strata indicates that this borehole is in the same limb of a major fold as the strata of Borehole C. However, since the two sequences of strata show no direct correlation it must be assumed that a fault occurs running between Boreholes C and D. (Naturally, there is insufficient evidence from just two boreholes to show which way the fault throws. However, we can give a minimum figure for the amount of throw.)

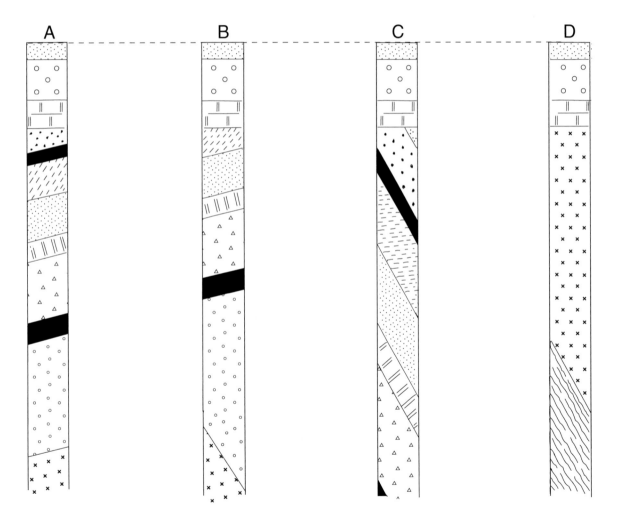

Igneous and impact features

Igneous features on geological maps can be sub-divided into volcanic rock units, covering both lavas and pyroclastic deposits, and intrusive rock units, which range in size from minor intrusions (e.g. sills and dykes) to large plutons (e.g. batho-liths and stocks).

Volcanic activity involves magma, usually a molten silicate melt, erupting at the surface of the earth (Plates 19 and 20). Intrusive activity, by con-trast, involves subterranean injections of magma into bodies of existing (country) rock. These can take place at different depths and give rise to a variety of intrusive types.

Although there are currently no active volca-noes in Britain, there are igneous features dating from the Precambrian to Tertiary, which are found mainly in the north and west of the country.

Lava flows

The term lava is used for the molten extrusive rock that flows from volcanoes onto the surface. They will bury existing rocks and may be subsequently buried by later lava flows, pyroclastic deposits or sediments. As a result they become interbedded with other strata, often conformably, and thus con-cordant with the strata above and below.

The form of any lava within the geological suc-cession is largely dependent on the viscosity of the magma as it is erupted. Basic and intermediate magmas, which produce basalt and andesite lavas respectively, are of lower viscosity, enabling them to flow over considerable distances. They are rarely more than 10 m in thickness but can cover many square kilometres in area.

Plate 19 The 1971 eruption of Mount Etna (Citelli vents), Sicily, Italy. Lava is flowing into a former river valley and filling up an existing topographic low.

Plate 20 Columnar Jointed Basalt Lava, Fingals Cave. This is a lava flow of early Tertiary age on the Scottish island of Staffa. Shrinkage during cooling of the lava has introduced polygonal columns at 90° to the bedding direction. The lava dips to the right and away from the camera.

More silicic, acid magmas tend to be highly viscous, which means that they form thicker lava flows which are much more limited in their aerial extent.

The viscous flow of such magmas is clearly recorded in the presence of flow banding (Plate 21) which records slight differences in the crystallinity and composition between individual layers of viscous magmas. The final cooling of the top surface of such lavas often shows a mass of angular, flow-banded blocks cemented together by the still molten lava as it moved slowly downhill. This feature is known as autobrecciation and is an important way-up criterion used by field geologists (see Chapter 6).

Another aspect of more viscous flows is that, unlike basaltic lavas, their internal flow banding may not be parallel to their contacts with older rocks. Changes of slope, ponding and the cessation of flow at the snout of the lava can all lead to upwarping due to compression and the formation of ramp structures (Plate 22). In extreme cases, these can be nearly vertical to the base of the lava, such as seen in Monte Amiata in North Central Italy and more locally in the Mew Stone Rhyolite

in the south of Skomer Island, Pembrokeshire, South West Wales.

Acid lavas of rhyolite, dacite or obsidian will therefore be formed in the central zones of any volcanic area, whereas more fluid lavas, usually of basalt or andesite, will flow off the central volcano or away from the eruptive fissure into areas of low relief. Here, the flows will be nearly horizontal and contacts close to topographic contours. Extensive piles of lavas can result in such areas, each separated by reddened zones of sub-aerial weathering and occasional lacustrine sediments. The differential weathering of the slow cooling central parts, which show the characteristic columnar jointing, and the intervening reddened palaeosols from the sub-aerial weathering of the lavas between eruptions produces a characteristic trap topography. This 'stepped' landscape is very characteristic of the Antrim Coast of Northern Ireland (Plate 23).

Although the topography above is solely attributable to lava flows and can be identified on a geological map in terms of contours and designated rock type, the concordant nature of many lavas means that they are difficult to distinguish from sills (see the section later in this chapter on

Plate 22 Ramp structure within phonolite lavas. Montana Rajada, Tenerife, Canary Islands.

Plate 21 Obsidian Lava Flow showing flow banding and flow folds. Colata delle Rocche Rosse, North East Lipari, Aeolian Islands, Italy.

Plate 23 Trap topography within series of basalt lavas, near Giant's Causeway, Co. Antrim, Northern Ireland. Note reddened Intra-Basaltic horizon (marked by arrow).

concordant intrusions) and are difficult to recognise from map evidence alone. The identification of rock type within the map's legend may also not help as minor intrusions can be composed of the same rock type (e.g. basalt).

However, field study of lava flows will clearly reveal the presence of a reddened, weathered upper surface (e.g. Skomer Volcanic Group at Marloes Sands, Pembrokeshire) and a single zone of baking (contact metamorphism) beneath the flow where it is in contact with older sediments. A concordant sill, by contrast, will show contact metamorphism both above and below and will not show any sub-aerial weathering zones.

Another aspect of basaltic and occasionally andesitic lavas is the formation of pillow lava (Plate 24) on eruption under water or on entry into water after a sub-aerial eruption. The interaction with heated seawater, in both cases, initiates metasomatism and a rise in sodium content. The resultant rock is designated a spilite and it is this term that would be the only clue to a submarine origin on a geological map. In the field, the fresh 'pillows' pile up around the advancing lava flow and, on compression, become squeezed into earlier still cooling pillow masses. This results in the structures seen in Fig. 11.1, with the pointed bases of the individual pillows facing downwards. This is one of a series of useful way-up criteria already discussed in Chapter 6.

Plate 24 Pillow Lava, Newborough Warren, Anglesey, North West Wales. Succession is overturned with older horizons to the left.

In terms of a geological map, especially in areas of more recent volcanism, the contacts between lavas and older rocks can be complex, especially if eruption has occurred onto an area of high relief close to the volcano (say within 5 km). In this zone, early erupted lavas will often be channelled down the volcano, filling pre-existing drainage channels (Plate 19). On solidification, they can therefore be exposed at a topographically lower level than the older lavas on which the drainage system has developed. However, since the lavas have erupted one after another onto the earth's surface, evidence of age will be easier to ascertain in more distal areas where the Principle of Superposition

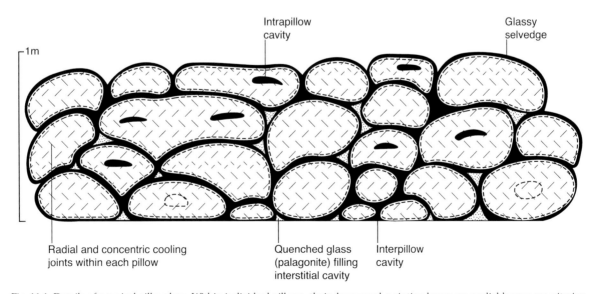

Fig. 11.1 Details of a typical pillow lava. Within individual pillows, their downward-pointing bases are a reliable way-up criterion.

can be applied. Indeed, on modern geological maps of Etna, many of the flows are dated, reflecting the long period of research and observation on Europe's largest volcano from the mid-thirteenth century to the present day.

Pyroclastic rocks

All volcanoes erupt both molten and solid pyroclastic material and indeed many of our island arc volcanoes, sitting above active subduction zones, often produce less than 10 per cent lava and over 90 per cent pyroclastics.

Pyroclastic rocks result from two main processes, namely pyroclastic fall and pyroclastic flow. Pyroclastic fall deposits (Plates 25 and 26) are produced by explosive eruptions of varying intensities at the volcanic centre and are dispersed downwind prior to falling back to earth to accumulate on land or to be redeposited through a body of water. Here it will be partially mixed with detrital sediment, giving rise to a volcaniclastic sediment. In mapping terms, these deposits can be treated in the same way as sedimentary strata. The finest material in such deposits can be transported by the wind for 1,000 km or more and can act as important datum levels within a complex series of sedimentary layers, especially if they are largely devoid of fossil assemblages. A good example is the Townsend Tuff Formation in the Lower Devonian, a pyroclastic fall deposit, which successfully correlates the diverse sequences of Old Red Sandstone across South Wales and the southern Welsh Marches.

Pyroclastic flows result from near-horizontal dispersal of a superheated aerosol of vesiculated lava blebs (pumice), solid fragments of older lavas and sedimentary rocks, and crystals all within a gaseous matrix. Temperatures at Mount St. Helens in 1981 reached 750°C and speeds of flow as high as 60 ms^{-1} have been recorded.

Many pyroclastic flows show detachment from their source vents and their deposits accumulate in valleys and depressions around the volcano, especially in its proximal (5–15 km) and distal (beyond 15 km) zones. The ash cloud above the moving pyroclastic flow is dispersed over a wider sector by the wind and will be deposited as a mantle across the whole landscape.

In some cases, the sudden emplacement of pyroclastic flows onto distal water-saturated mud-

Plate 25 Bedded pyroclastic fall deposits, South East Lipari, Aeolian Islands, Italy. An alternation of coarse and fine pumice deposits derived from pulsating eruptions. Channelling at the base of some of the coarser units indicates that the sequence is right way up.

flats causes minor tremors, sediment liquefaction and the collapse of the ash flow into the sediment. Some of the resultant load casts at the base of the flow become detached and end up as isolated steep-sided bodies, totally enclosed by the sedimentary layers (Fig. 11.2). An Ordovician example of this process is found within the Capel Curig volcanic formation in North Wales on the southeastern edges of Snowdonia.

Evidence from older Quaternary deposits suggests that pyroclastic flows are capable of travelling over 100 km from the vent and at speeds of >100 ms^{-1}. At this speed, such flows can climb uphill and totally mantle the landscape. The consolidation of such pumice-rich pyroclastic flows produces ignimbrite as a rock unit. There is little observational information on such huge eruptions, the eruption in the Valley of Ten Thousand Smokes in Alaska in 1912 perhaps coming nearest to these gargantuan events. Certainly, in the Ordovician of North Wales, huge thicknesses of ignimbrite, showing the high levels of internal welding associ-

Plate 26 Unconformity between two distinct phases of pyroclastic fall deposits near an observatory, South Lipari, Aeolian Islands, Italy.

ated with the elevated temperatures of emplacement, have been erupted. These can be mapped as single events (Pitt's Head Ignimbrite) or, in combination with detailed field observations, a series of individual flows (Lower Rhyolite Tuffs). In this sense, they can be treated in the same way as lava flows.

Concordant intrusions

Sills

The majority of these are sills composed of basalt, dolerite or felsite. Sills are sheet-like bodies of magma, ranging from millimetres to many metres in thickness, injected between pre-existing beds of

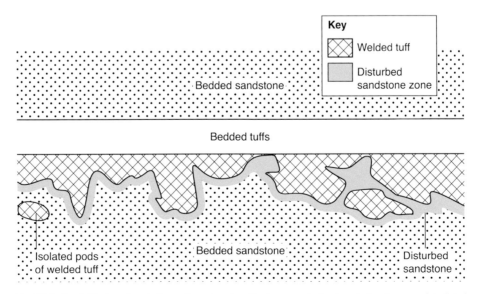

Fig. 11.2 Capel Curig volcanic formation showing the large-scale load cast structures and isolated pods of welded tuffs at its base. This is another example of a way-up criterion.

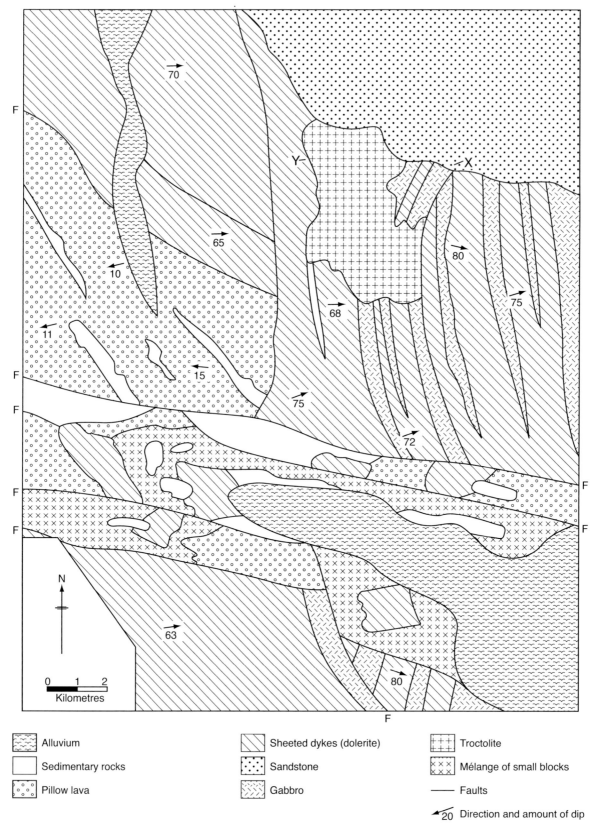

Map 30 Ophiolite complex in Oman. This map has been modified from that of part of Masirah Island, Oman and represents a section of oceanic crust uplifted above sea level and extensively eroded.

Legend

- Alluvium
- Sedimentary rocks
- Pillow lava
- Sheeted dykes (dolerite)
- Sandstone
- Gabbro
- Troctolite
- Mélange of small blocks
- Faults
- Direction and amount of dip

Questions on Map 30

The sheeted dyke complex varies from 100 per cent dykes to dykes with variable percentages of gabbro or pillow lava screens.

Study the map and familiarise yourself with the key provided.

1. The sheeted dyke complex in both the northern and southern sectors of the map trends north–south. In which direction and at what angles do the dykes dip?
2. Which rock type outcrops to the west of the dykes? What is its dip direction?
3. Which rock type outcrops to the east of the dykes? What is its angle of dip?
4. Which rock type is restricted to the prominent east–west trending structure that cuts through the dyke complex?
5. Describe the orientation of the fragments of sheeted dyke complex in this east–west zone.
6. Identify the nature of geological contacts X and Y.
7. From your knowledge of the geological conditions prevailing at mid-ocean ridges, identify:
   a) an uplifted sea floor fragment
   b) the direction on the map pointing towards deeper crustal levels
   c) an uplifted transform fault system
   d) the probable directions of former sea-floor spreading processes at this locality.

disturbance. Of course, faulting that post-dates both the strata and the sill will displace both.

A sill is generally concordant with surrounding beds but can change its horizon, appearing between different beds in different places. Often, such changes in stratigraphic position occur in abrupt steps because the magma has passed from one level to another using a joint or fault (Fig. 11.3). This can be observed in the Great Whin Sill of northern England.

The Whin Sill also shows another sill feature: changes in thickness. It varies in thickness from 30 m to 39 m in various places. A sill may also split into several 'leaves' and this can be seen in the Tertiary sills on BGS Map 80: Trotternish Peninsula of the Isle of Skye. This phenomenon is not typical of lava flows although they may have smaller-scale rafts of sedimentary rock caught up within them.

Composite sills (and composite dykes) occur when there is more than one injection of magma. The first usually forms the outer margins of the sill and a second, later, intrusion often runs through the centre, sandwiched by the earlier, solidified, magma. It is possible to see this on larger-scale geological maps.

Sills are composed of resistant rock so they often cap hills, such as those of the Fife–Kinross border, protecting the softer rocks beneath. A sill may also form a pronounced escarpment, leading to waterfalls and providing a natural defence feature. Hadrian's Wall is partly constructed along the

rock (Plate 27). They run parallel to the bedding direction and are said to be concordant. The intrusion seldom causes any observable disturbance of the strata so that a sill behaves structurally as though it were part of the stratigraphic succession. If there is subsequent tilting and folding of the strata, the sill is tilted and folded along with everything else. Sills in steeply dipping strata could be confused with dykes.

Using a map, a sill can be distinguished from surrounding strata in a few respects. It will, of course, have a different petrology to the surrounding rocks. It has also been intruded some time after the surrounding strata have been laid down. The strata may have already been displaced by faulting but the later sill may pass across faults without

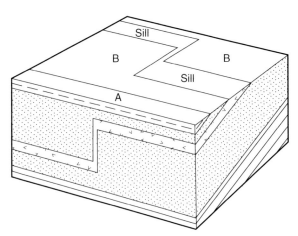

Fig. 11.3 Block diagram showing a sill intruded into dipping strata. Note that the sill is seen to change horizon abruptly.

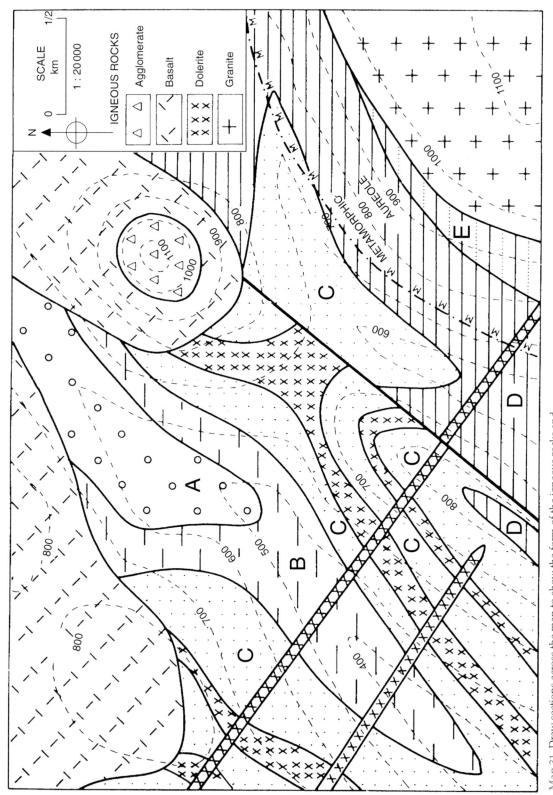

Map 31 Draw sections across the map to show the form of the igneous rocks and the other structural features. Also deduce the relative ages of the igneous rocks as far as this is possible.

Plate 27 The Drumadoon Sill, Isle of Arran, Scotland showing clearly developed columnar jointing.

Whin Sill and Stirling Castle sits upon a sill escarpment.

Other concordant intrusions, such as laccoliths and lopoliths, are rare and not covered here.

Discordant intrusions

Dykes

These are vertical or steeply dipping sheet-like intrusions, from a few millimetres to several metres thick (rare examples reach tens of metres across). They can often be traced across country for many kilometres.

Since they are approximately vertical, dykes follow straight paths across the landscape regardless of changes in topography. It is sometimes possible to date a dyke relative to the surrounding sedimentary rocks. It must be later than the youngest beds through which it cuts and, if it lies beneath an unconformity, will be older than the beds above the unconformity if it does not cut through these beds (cf. the dating of faults, p. 71).

Dykes are intruded during periods of crustal tension. Widening cracks or joints permit magma to be intruded both from below and laterally. Dykes may be very numerous near to volcanic centres (Plate 28), where they may indicate a considerable extension of the crust. The dyke swarm on the south coast of the Isle of Arran is a good example (Map 35).

At mid-ocean ridges, sheeted dyke complexes represent the succession of fissures which once fed ocean-floor basalt eruptions of pillow lava. Their orientation is a direct result of sea-floor spreading processes and at deeper levels in the crust they become more gabbroic (coarser grained) as they approach the mid-ocean ridge magma chambers. A typical uplifted portion of ocean crust in Oman is

Plate 28 A series of composite dyke intrusions on the road to Il Portillo, Tenerife, Canary Islands.

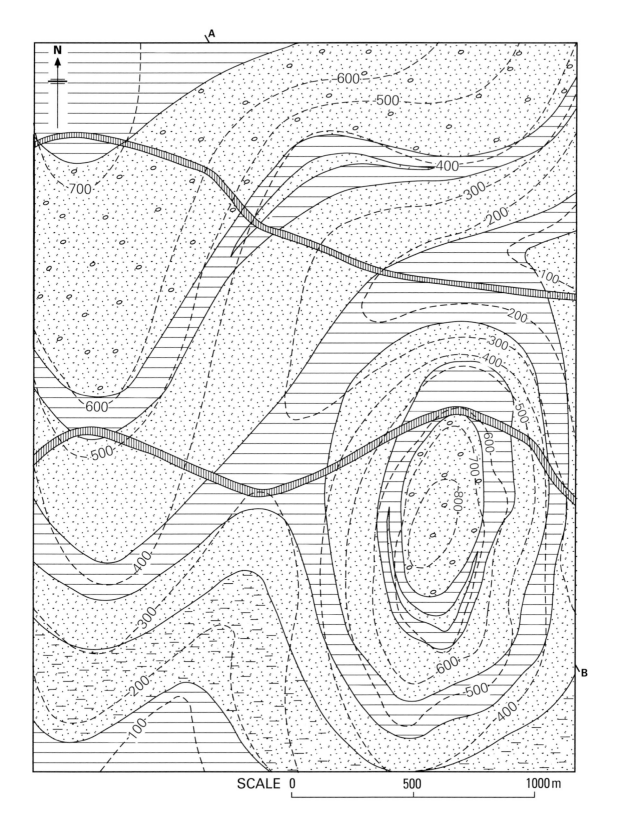

SCALE 0   500   1000 m

Map 32 How many dykes and sills are visible on this map? (Dykes are vertically shaded and sills are horizontally shaded.) What criterion have you used to distinguish the sills from the dykes? What evidence is there against the sills being lava flows? Draw a section from A to B using the profile provided. Calculate and comment on the dip of the dykes.

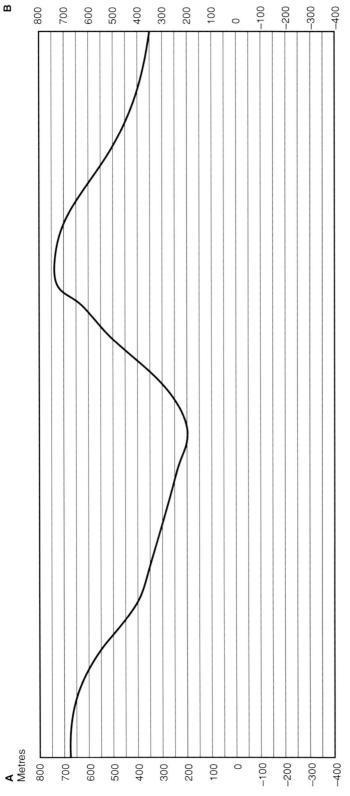

Fig. 11.4 Topographic profile of section A–B, Map 32.

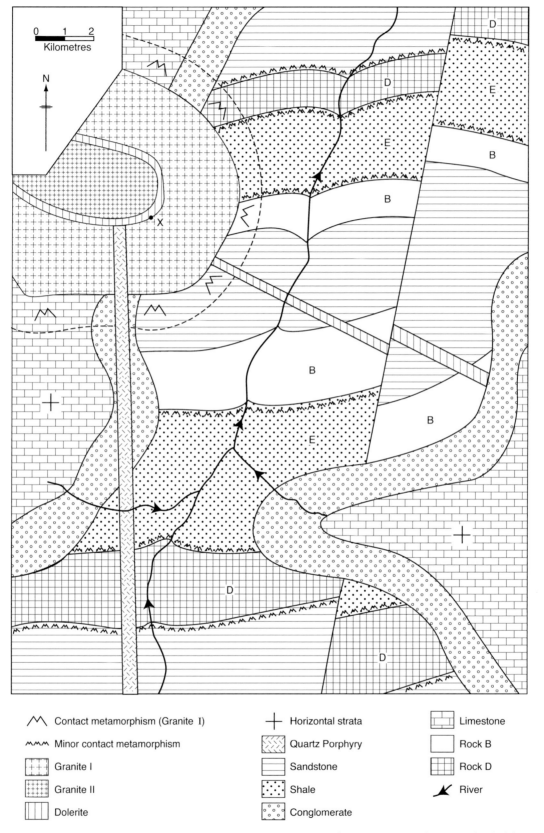

Map 33 This geological map contains a wide range of both folded and horizontal sedimentary strata together with igneous intrusions of varying types and age. A detailed key to the various rock types is provided.

Key

∧∧ Contact metamorphism (Granite I)

∧∧∧ Minor contact metamorphism

Granite I

Granite II

Dolerite

+ Horizontal strata

Quartz Porphyry

Sandstone

Shale

Conglomerate

Limestone

Rock B

Rock D

River

seen in Map 30. The whole suite of pillow lavas, dykes and gabbros together with ocean-floor pelagic sediments is termed an ophiolite complex.

Although most dykes are essentially linear features, those related to central igneous complexes are characterised by sheet-like intrusions that appear as circular rather than linear outcrops on geological maps. These are termed ring intrusions and are classified in two types.

Ring dykes

Ring dykes display vertical or steep outwardly dipping contacts and have been emplaced by the subsidence, often in several stages, of a central cylindrical crustal block, in a tensional stress field. Magma, sometimes in multiple injections, has then risen to fill the void created. As can be seen in Fig. 11.6, the level of erosion can involve very different outcrop patterns on a geological map. At the lower erosion level, a ring dyke as described above outcrops with sedimentary rocks at its centre. With less erosion, a circular body of igneous rock or concentric circles of igneous rocks will result. These can easily be confused with steep-sided plutonic intrusions (see later). Overall, a ring complex can often be between 5–10 km in diameter made up of multiple intrusions.

Cone sheets

These are also circular in plan and form continuous or discontinuous arcs which dip steeply inwards towards the volcanic centre. In contrast to ring dykes, they are formed by the uplift of a central conical block by the intruding magma and its entry into the multiple tension fractures formed by this process. They occur in swarms and a notable example can be seen on the Ardnamurchan peninsula in Scotland. Cone sheets are generally thinner than ring dykes, being only a metre or two in thickness.

Stocks, cupolas and batholiths

These are larger plutonic intrusions, often of granite or granodiorite, emplaced deeper within the crust. They become exposed by the prolonged erosion of the overlying cover of rocks. Near-circular stocks and irregularly shaped bosses are usually tens of kilometres across and represent a single intrusive

event. Batholiths are larger intrusions, which can reach hundreds of kilometres in size and usually involved multiple intrusion episodes.

Many batholiths show an irregular upper surface with magmatic extensions towards the surface, called cupolas, separated by remnants of the former sedimentary cover, called roof pendants. Dartmoor Granite is one such cupola and is part of the Cornubian batholith, which extends under much of the South-West Peninsula.

The intrusion of granite in the upper levels of the crust involves the penetration of the magma along bedding planes and joints within the enclosing country rocks. This process detaches blocks of the country rocks, which then sink into the magma and are eventually assimilated into the molten melt. This is the process of piecemeal stoping and it gives rise to xenoliths of country rock, in various states of assimilation, within the margins of granite plutons.

These discordant intrusions are often longer than they are broad because they follow the structural geology of the region, but in every case they have steep margins, which cut through the surrounding and pre-existing strata. Because they are composed of relatively resistant rock, large intru-

Questions on Map 33

1. Identify and mark the following features:
   a) an angular unconformity
   b) a sill
   c) a lava flow
   d) an area of horizontal rock strata
   e) a dyke intruded earlier than Granite I
   f) an axial trace of a major fold (state type).
2. Mark on the map the downthrow side of the major NNE–SSW fault.
3. Assuming that Bed B is a constant thickness in its outcrops within the river valley, in which outcrop is its dip steeper and in what direction?
4. Comment on the geological history of the granite outcrops in the north-west of the map and especially the dolerite structure seen at Locality X.

Outline the geological history of the map area as a series of numbered statements, starting with the oldest or earliest event.

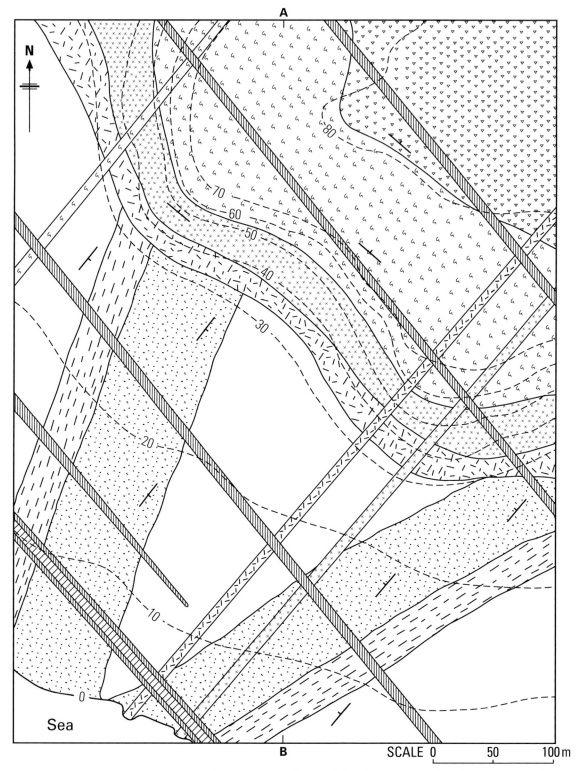

Map 34 How many igneous episodes are indicated on this map? (As well as the dykes, all units above the 30 m contour are igneous.) To which episode would you assign the composite dyke? Indicate an unconformable boundary on the map (other than dyke boundaries). Draw a section from A to B. Calculate the dip of the lavas and tuffs. What is the probable dip of the dykes?

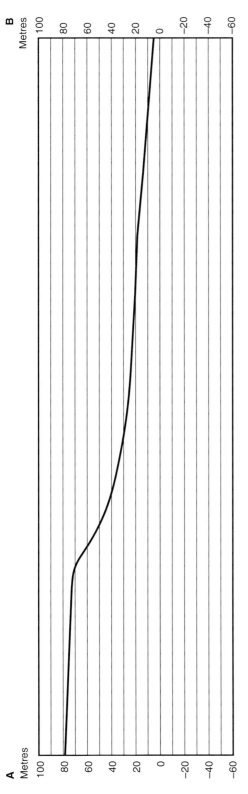

Fig. 11.5 Topographic profile of section A–B, Map 34.

sions can form upland areas. They are surrounded by a zone of rocks that have been altered by heat from the magma. These contact metamorphic aureoles are typically a few kilometres across at most. Smaller intrusions, such as sills and dykes, affect only a few centimetres or metres of the rock surrounding them. Large intrusions also cause extensive mineralisation. Mineral veins are often dyke-like on maps because they fill vertical fissures. (Block Models 9 and 10, Appendix.)

Volcanic necks

These result from the infilling of volcanic vents with consolidated lava (basalt, etc.) or pyroclastic material (vent agglomerate, tuff, etc.). They cut through existing strata, have near-vertical sides, and they are usually circular in plan. A volcanic neck has the same shape as a boss but is much smaller and is composed of different rocks.

Using a map, it is sometimes difficult to distinguish between a volcanic neck and a near-circular hill-top cap of igneous rock (a sill or lava flow). In section, however, they look very different (Fig. 11.7).

Impact features

During the development of the early Solar System, there were numerous impacts of fast-moving comets and asteroids onto the surfaces of the newly forming planets. As the planets swept clean their orbits, the number of larger objects capable of potential collision subsequently decreased. Thus, on any particular body, the greater the crater density, the older the surface. Callisto, one of the Galilean satellites of Jupiter, displays such an ancient

Exercise on geological survey map

1. **Edinburgh: 10 (Map 32, Scotland).** Give the relative ages of the various types of igneous intrusion shown on the map. How can the relative age of an igneous rock be discovered from map evidence and what are the limitations of this method? What other evidence might one expect to find in the field but which cannot be discovered from a map? (This map has now been replaced by 32E Edinburgh and 32W Livingstone on the 1:50,000 scale.)

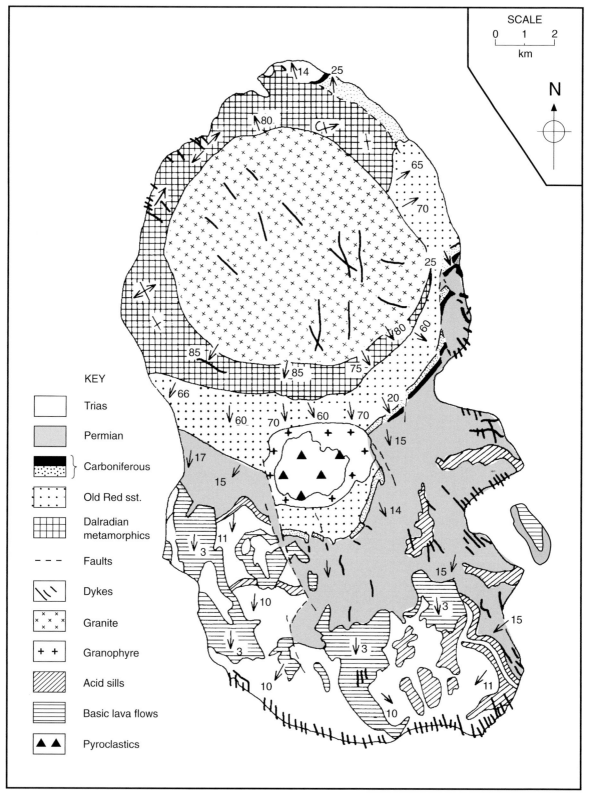

SCALE

0   1   2
km

N

KEY

	Trias
	Permian
	Carboniferous
	Old Red sst.
	Dalradian metamorphics
- - -	Faults
	Dykes
	Granite
+ +	Granophyre
	Acid sills
	Basic lava flows
▲ ▲	Pyroclastics

Map 35 This map is broadly adapted from the geological map of the Isle of Arran 1:50,000 Special Sheet. Most of the kinds of igneous intrusion shown on the stylised Map 33 can be found here. Write an account of the geology of the area, noting particularly the two major unconformities, the dips of strata of different ages and the distribution and disposition of different igneous rocks. What can you deduce from the general direction of the dykes? It is possible to work out the geological history for this map (however, it will not be identical to that of the one-inch-to-the-mile Geological Survey sheet). Reproduced by permission of the Director, British Geological Survey: NERC copyright reserved.

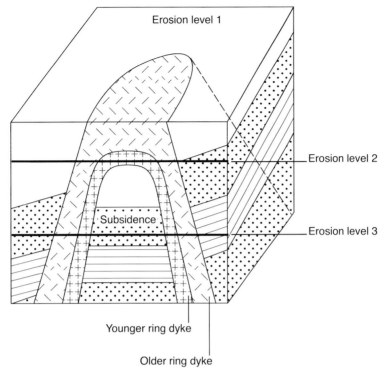

Fig. 11.6 Block diagram showing ring dyke formation and the effect of different erosion on their appearance on maps.

surface at least 4,000 Ma old. Continual impact processes can rework surfaces so that craters are gradually obliterated or become 'ghost' craters.

As on earth, the principle of cross-cutting features, as used in the study of volcanic centres, can be applied to any planetary surface and deductions about the ages of the various events can be successfully made from a distance.

Like other rocky planets in the Solar System, the earth itself has been bombarded by impacts (Plate 29) but due to continuing plate tectonic activity and the weathering of its surface by water, wind and ice, the detection of these structures (or 'astroblemes') is more problematical. Typical of impact structures on the earth is the Ries Crater in southern Germany. It is 25 km in diameter and, as with other larger meteoritic impact craters, is a double-ringed circular basin. The basin is now occupied by the town of Nordlingen and several villages set in a rural landscape. Exposures are

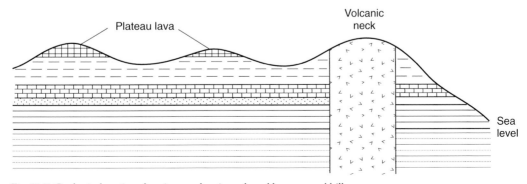

Fig. 11.7 Geological section showing a volcanic neck and lava-capped hills.

The diagram represents a quarry face that exposes a variety of igneous and sedimentary rocks. By studying the field relationships between the various rock units, answer the following questions.

1. Give, with reasons, the names of the igneous bodies represented by X, Y, Z.
2. Using the radiometric dates determined for the igneous rocks and evidence shown in the quarry face, designate the geological age for sedimentary rocks A, B, C and D.
3. Comment on the evidence for the structural orientation of pillow lava B.
4. Igneous Rock Z shows extensive flow folding and autobrecciation. Comment on its likely composition.
5. Identify the nature of Rock E.
6. State the nature of the boundary N–N, giving two reasons.
7. Outline, in a series of statements, the geological history of the area (oldest first).

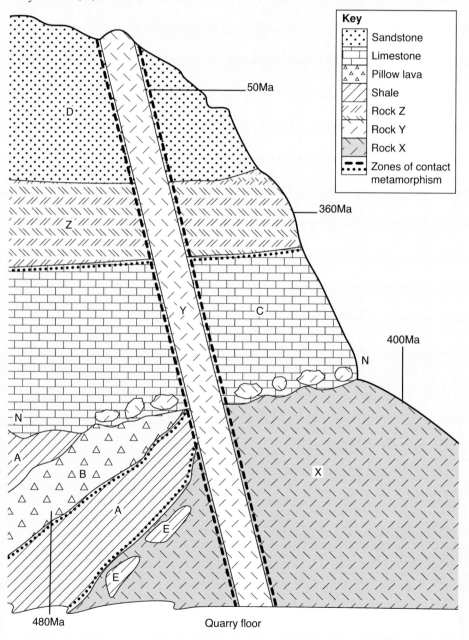

Key

(dotted pattern)	Sandstone
(horizontal lines)	Limestone
(triangle pattern)	Pillow lava
(hatched pattern)	Shale
(double-line pattern)	Rock Z
(single-line pattern)	Rock Y
(dash pattern)	Rock X
▬ ▪▪▪	Zones of contact metamorphism

50Ma

360Ma

400Ma

480Ma

Quarry floor

Plate 29 A view across Meteor Crater, Arizona, USA. This 50,000-year-old impact feature is surrounded by extensive deposits of ejecta débris.

relatively rare but its crater morphology, structure and ejecta deposits are very well preserved. Its age has been put as Middle Miocene (14.7 Ma).

The impacting body, estimated at between 800–1200 m in diameter and travelling at 20–60 km sec^{-1}, produced two major geological deposits. Firstly, as the asteroid penetrated to the metamorphic basement, volatilised and exploded, the surface was opened up and the Bunte Breccia was ejected. This is largely made up of Triassic and Jurassic sedimentary rocks with a smaller and early component of basement crystalline fragments. This ejecta sheet extended out to 37 km from the crater in the south-west and contains rock masses up to 1 km in size (schollen) which are large enough to feature on the local geological maps. At the same time as the Bunte Breccia was ejected sideways from the developing impact crater, a 600°C impact melt containing rock fragments and mineral fragments was ejected vertically into the stratosphere. This molten melt eventually returned to earth and mantled deposits, including the ejecta sheet of Bunte Breccia, not only in the crater but outside the crater to a distance of 32 km (Plate 30). This melt on solidification is known as suevite and has a distinctive texture of elongate glassy fragments and blebs. The Romans found it to be an excellent

Plate 30 Red Bunte Breccia overlain by suevite. Otting Quarry, near Nordlingen, Bavaria, Germany. (*Photograph by Diana Smith*)

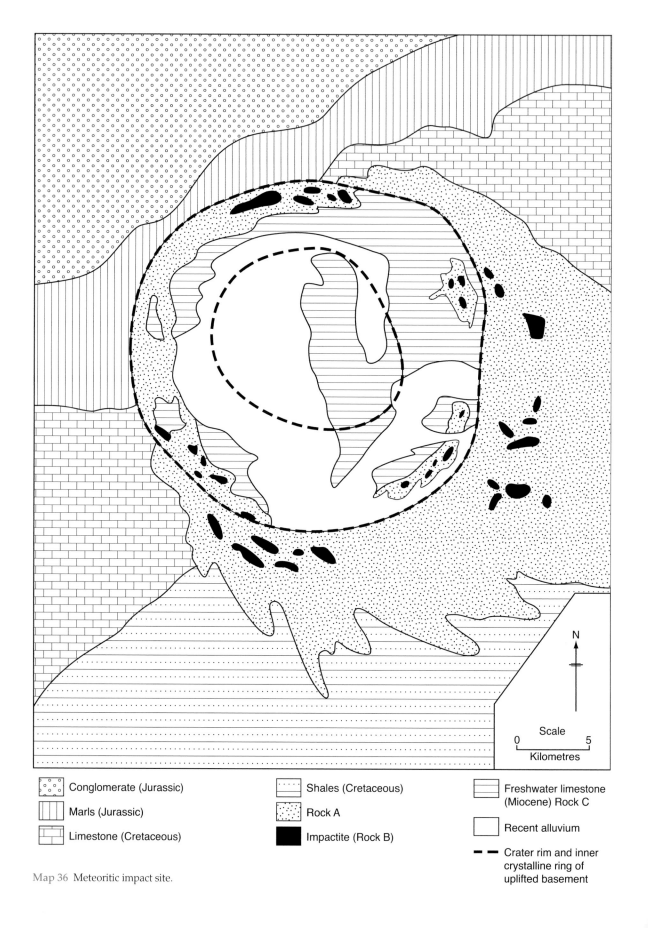

Map 36 Meteoritic impact site.

Conglomerate (Jurassic)

Marls (Jurassic)

Limestone (Cretaceous)

Shales (Cretaceous)

Rock A

Impactite (Rock B)

Freshwater limestone
(Miocene) Rock C

Recent alluvium

Crater rim and inner
crystalline ring of
uplifted basement

N

Scale

0   5

Kilometres

Questions on Map 36

1. Using the key which describes the rock types and their stratigraphic ages, identify:
   a) the strike of the pre-existing rocks
   b) the centre of impact
   c) the ejecta blanket
   d) the arrival direction of the impacting body.
2. Give an explanation for the outcrop pattern of Rock C.
3. Calculate a possible age for the impact crater.

building stone and St George's Church in Nordlingen is totally constructed of the material.

On a geological map, the more recent impact craters are usually circular or near-circular structures with distinctive rim (or double rim). They bear no relationship to the strike of the surrounding sedimentary rock types and often contain deposits which are physically constrained by the crater walls (Map 36). Breccias and impactites (impact melt) can be identified within the local stratigraphic column.

Our overall knowledge of impact cratering on the earth now parallels the current growth and quality of space probe imaging. Geological mapping principles can now be applied not only to the surfaces of the inner rocky planets but also to the icy moons of the outer planets.

Unlike the earth, many solar system bodies have largely unmodified surfaces showing ancient impact structures. Craters are gradually obliterated on bodies with little or no atmosphere by continual micrometeorite impacts or by resurfacing due to volcanic activity (e.g. Jovian moon Io with its active eruptions of sulphur lavas). Planets and moons such as Titan which have substantial atmospheres experience surface erosion due to three processes, namely the action of wind, the movement of water or, in the case of Titan, methane and the slow creep of ices of various compositions upon their surfaces.

Map 37 shows a simplified map of part of the surface of Ganymede, one of the major satellites of Jupiter, and the Solar System's largest moon with a diameter of 5,262 km. A series of probes has resulted in excellent photographic coverage of its surface, which can be analysed geologically. With virtually no atmosphere, surface processes are minimal.

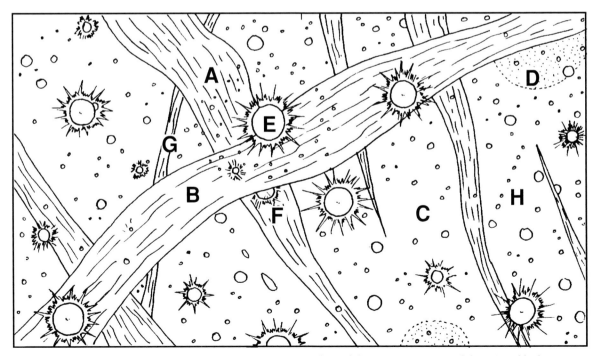

Map 37 The surface of Ganymede is represented by a line drawing. The surface is composed of ice which has become darkened by contaminants over time. The linear features are paler and the youngest craters and their ejecta blankets appear white.

Questions on Map 37

1. There are craters of various sizes. Suggest factors that will influence their dimensions.

2. Some of the craters are surrounded by ejecta blankets. What factors will influence the extent of these blankets? Why is the ejecta material white and why are these craters paler than the others?

3. Why is crater E surrounded by a 'halo' of smaller craters?

4. Suggest a cause for the linear pattern of craters at H.

5. The circular feature D has largely disappeared. What might it have been and what may have happened to it since? Refer to the next question for a clue.

6. The large linear features are cracks between crustal blocks. Suggest what has happened to the crust and a mechanism for it.

7. Of its four larger moons, Ganymede is the third most distant from Jupiter. The second moon, Europa, has a cracked icy surface with very few craters, whereas Callisto has no cracking but extremely dense cratering. Suggest how their surfaces have evolved compared with Ganymede and place the surfaces of these three moons in age order.

8. Arrange the surface features A to F in age order using the principle of cross-cutting.

12

Economic problems

In a book focused largely on map problems and their solutions, many aspects of economic geology and geotechnics must be left to other textbooks. However, the reader should appreciate that problems in both these fields are dependent upon the analysis of geological structures and in many cases economic calculations and design decisions are made on data derived from geological maps. We have seen the simplest application of an economic problem already in Map 10, the deduction of the position of the sub-unconformity 'outcrop' of a coal seam from structure contour intersections. Simple economic problems were also posed on Maps 20 and 21.

The six maps that follow illustrate how structural geology can be usefully applied to the exploitation of coal seams, ironstone horizons and ore bodies and to the building of tunnels for a hydro-electric scheme. The problems include the drawing structure contours and overburden isopachytes.

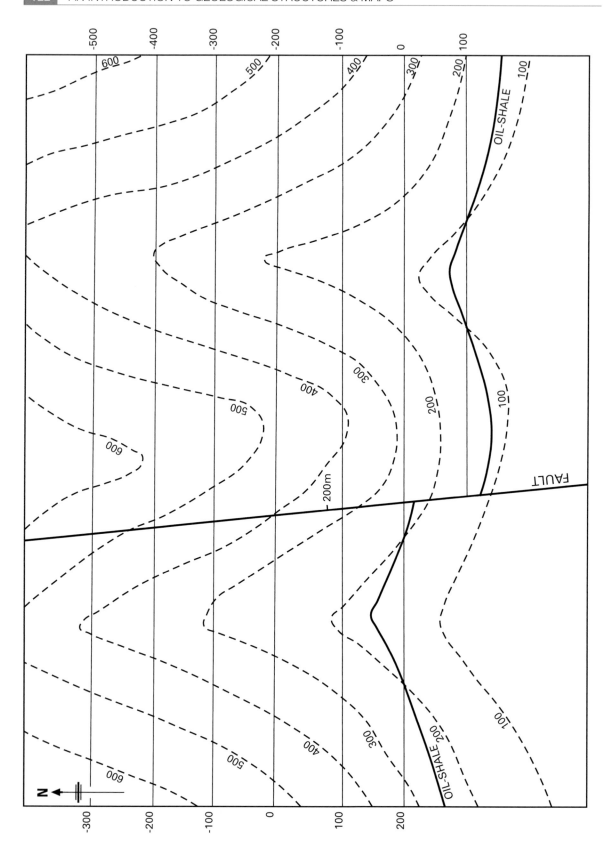

Map 38 The map shows an area of considerable relief with contours at 100 m intervals shown by broken lines. A thin but persistent oil shale outcrops and is cut by a dip fault with a downthrow of 200 m to the east. The structure contours for the oil shale, deduced from dip measurements on the outcrop and confirmed by borehole information, are shown by fine lines. Draw the isopachytes for the cover (= overburden) overlying the oil shale at 100 m intervals (100 m, 200 m, 300 m, etc.). Re-member that an isopachyte joins up points of equal thickness. The difference between the height of the ground and the height (or depth) of the oil shale at any intersection of the ground contours and the structure contours will give the thickness of the overburden at that point. All figures on the map are given as heights in metres relative to sea level. Economic considerations limit overburden removal to a maximum of 200 m. Indicate the area that can be profitably opencasted.

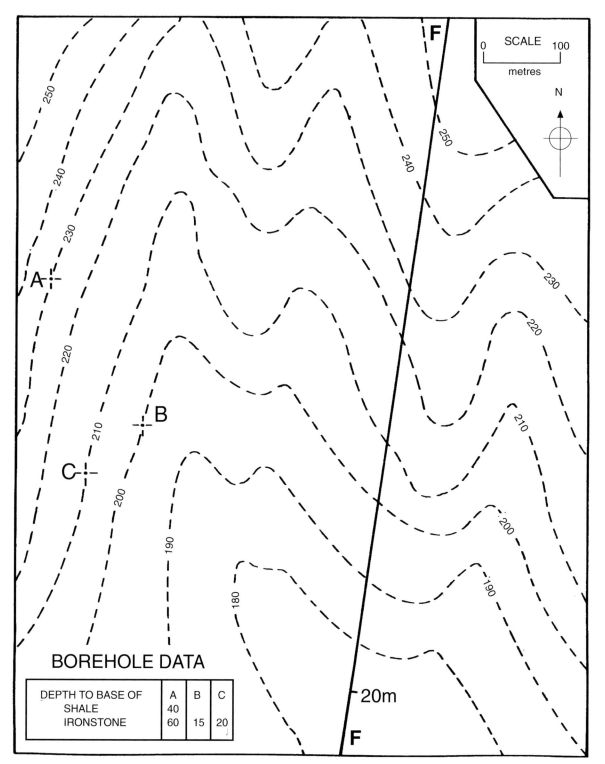

Map 39 The western part of the map comprises a three-point problem enabling us to draw structure contours at 10 m intervals on the base of the ironstone. Assuming the top of the ironstone to be 20 m higher, why does Borehole B penetrate only 15 m of ironstone? East of the fault (a normal fault of high angle of dip), produce the structure contours and renumber them 20 m lower (since the fault is shown as having a downthrow of 20 m to the east). Shade outcrops of iron ore on both sides of the fault. Also shade areas where the ironstone could be worked opencast if not more than 40 m of overlying shale is to be removed, i.e. draw the overburden isopachyte for 40 m. Draw a section along the north–south line N–S passing through point B and another section along the east–west line E–W passing through B.

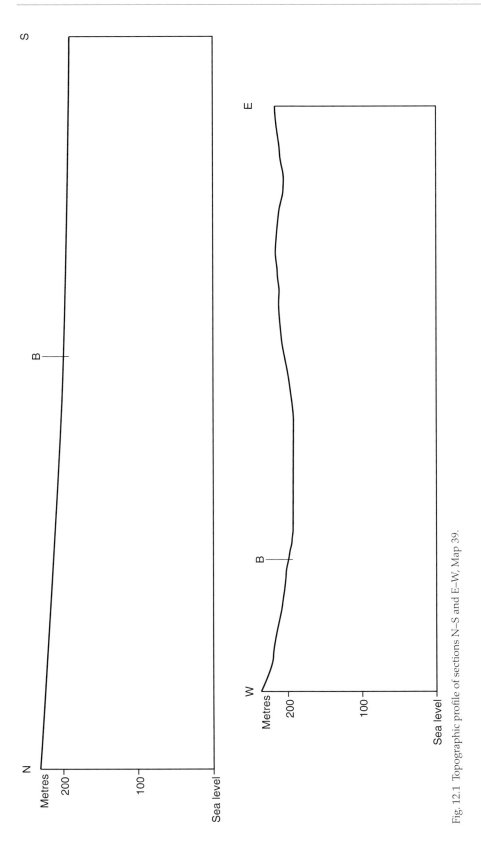

Fig. 12.1 Topographic profile of sections N–S and E–W, Map 39.

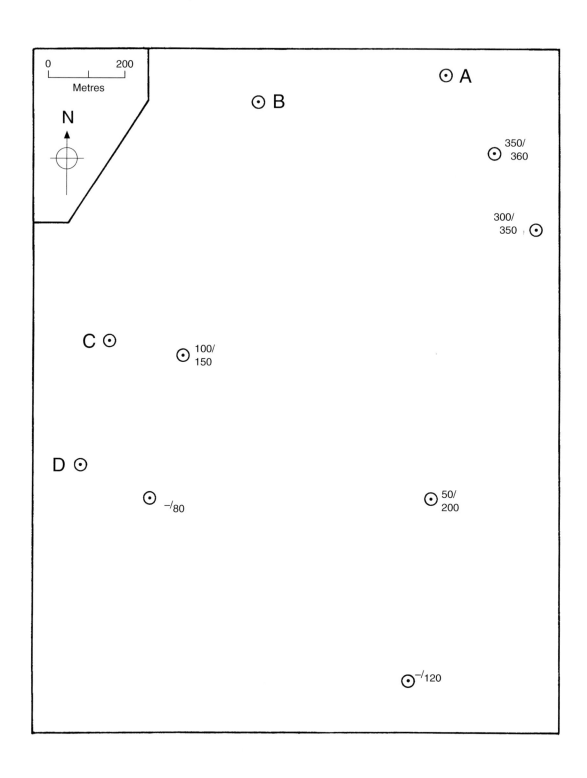

Map 40 A wedge-shaped, igneous, economically valuable sill-like body is encountered in boreholes. The land surface is flat and horizontal over this area and the depths from the surface to the top and base of the sill are given beside each borehole. Construct structure contours for the top of the sill and for its base, assuming that both top and bottom are uniformly dip-ping (but not parallel). Construct isopachytes at 50 m intervals to show how the sill varies in thickness. Shade the area where the sill outcrops. Why is the sill absent in Boreholes A and B? If it was not found in Boreholes at C and D, how would you explain its absence?

Note on Map 40. The reason for the absence of the sill in some boreholes could be as follows: (a) it outcrops so that at the site of the borehole it is no longer found as a result of erosion, (b) the sill thins to nothing and is not found, (c) if the absence of the sill is not due to its being eroded away or to its thinning, it must be absent in a borehole due to faulting. (The sill may have been upfaulted then removed by erosion or it may have been downfaulted to such a depth that the borehole was not deep enough to find it. There is no map evidence given here to indicate which explanation is the more probable.)

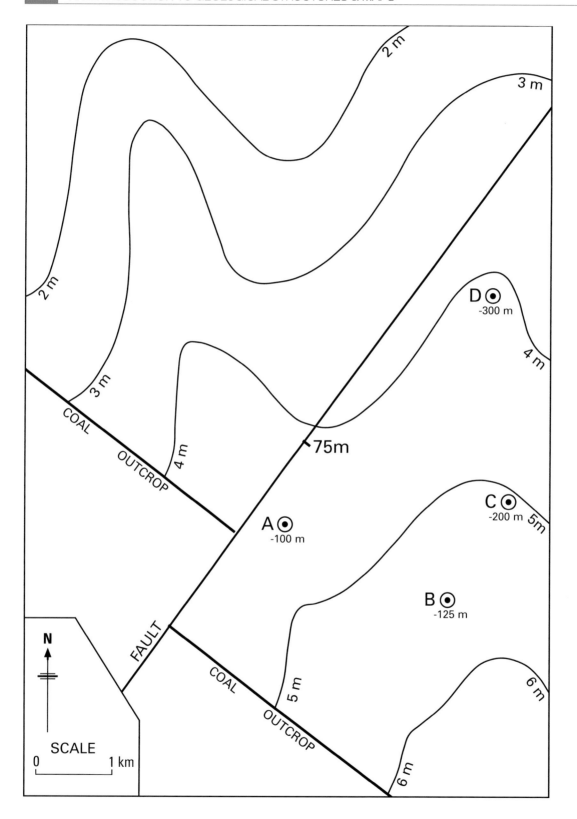

Map 41 The outcrops of a coal seam have been revealed in an area of flat ground. It can be seen that the lateral displacement of the outcrop is a consequence of a normal fault with a 75 m throw. The seam was encountered in four boreholes, A to D, at depths given on the map showing that it has a uniform dip in a generally north-easterly direction (37°E of N). The coal thins from a maximum of over 6 m in the south-east to less than 2 m in the north-west. The isopachytes on the map show that this thinning is not uniform. It is proposed that it would be economical to mine the coal by opencast methods (= strip mine) to a depth of 150 m where the seam is not less than 3 m thick. Draw structure contours on the top surface of the coal seam at intervals of 50 m and shade on the map the areas where the coal may be economically mined.

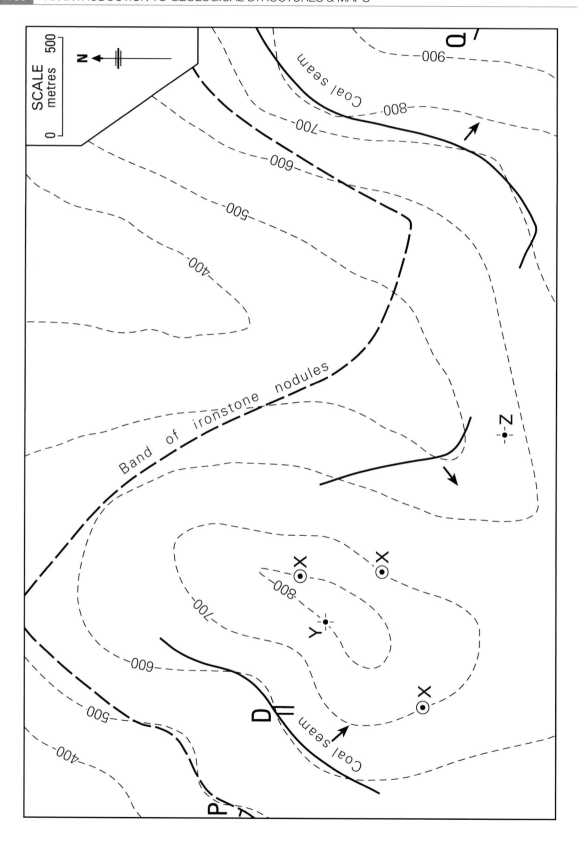

Map 42 The map shows part of the outcrop of a coal seam. It was also found at a depth of 300 m in the boreholes at several points marked X. Complete the outcrop of the coal seam. Determine the depth at which it would be encountered in shafts put down at Y and Z. Also insert the outcrop of another seam that lies 300 m higher (vertically) in the succession. What is the amount of plunge of the fold axes? Insert outcrops of the fold axial planes on the map. Draw a section along the line P–Q. Suppose an adit (drift mine) was situated at D. Naturally, this tunnel follows the slope of the coal seam, although it is running due south. What would be the gradient of this tunnel? Note: What you are calculating is the apparent dip of the coal on bearing 180° (expressed as a gradient).

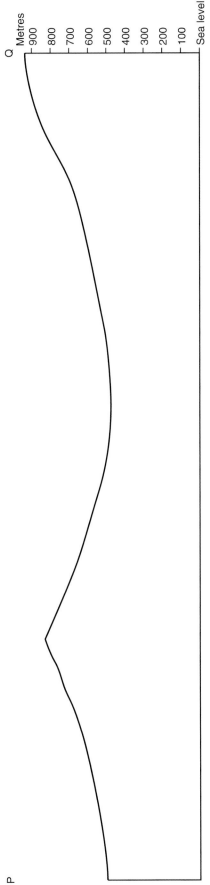

Fig. 12.2 Topographic profile of section P–Q, Map 42.

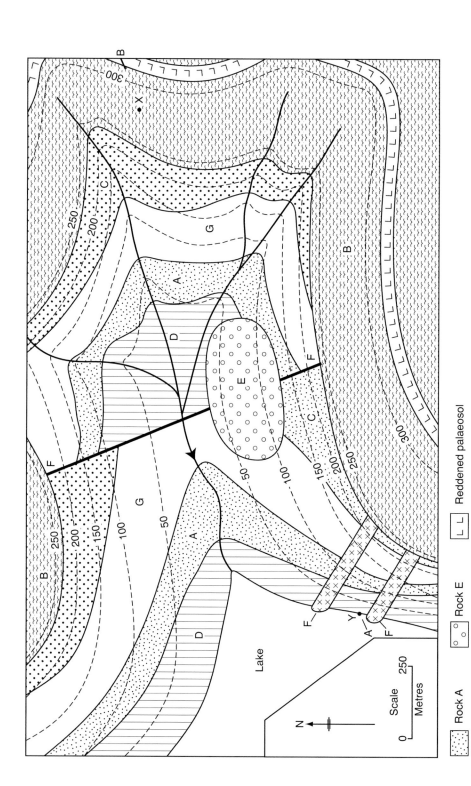

NB: Strata are not in stratigraphic order

Map 43 A new hydro-electric scheme requires a tunnel. As part of the site investigation engineering geologists have mapped in detail the area under consideration as shown in this map.

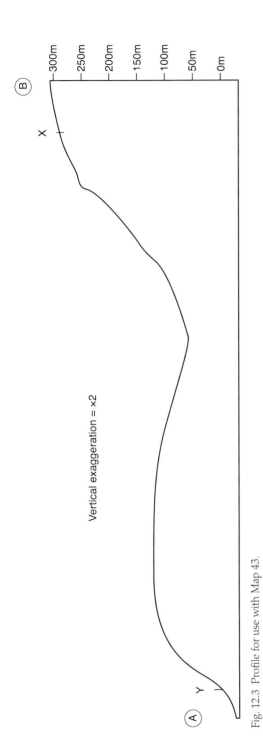

Fig. 12.3 Profile for use with Map 43.

Vertical exaggeration = ×2

1. After consultation with staff geologists, it has been decided to commence the tunnel at Point X (marked on the map and on the profile). A **vertical** shaft is to be driven down to sea level at this point before the tunnel is driven **horizontally** along the line of section to enter the lake at Point Y. Mark the shaft and the tunnel clearly as lines on the profile provided.

2. With careful study of the map, complete a geological section along A–B using the profile provided.

3. Through how many metres of Rock A, measured horizontally, will the horizontal section of the tunnel pass?

4. How many times will Rock D be seen in the horizontal section of the tunnel?

5. Rock G contains great volumes of water, which will be released by tunnelling. None of the other rocks on the map contains any water. Mark clearly on your cross-section any parts of the tunnel, in both the horizontal and vertical sections, where flooding may be expected.

6. At what depth would the **horizontal** section of the tunnel have to be placed to avoid Rock G?

7. What is the angle of dip of Rock G? What is the throw of the fault in metres? How can you tell from the map that the fault plane is vertical?

8. Using information provided on the key, identify the three igneous features represented by rocks B, E and F.

13

Complex structures

Nappes

Following the discussion in Chapter 6 in which overfolds were described, i.e. structures in which beds are overturned beyond the vertical, we now consider nappe structures. A nappe arises from a very large overfold in which the strata are nearly horizontal over wide areas. The fold structure is referred to as a recumbent fold and has been 'pushed over' so far that both limbs have low angles of dip, and are approximately parallel, although in the case of one limb the beds are actually upside down, i.e. the succession is inverted. Only at the 'nose' of the structure where the strata are folded back on themselves will steep dips be encountered (Fig. 13.1).

Of course, minor folding is usually superimposed on a major fold such as this so that locally steep dips may be seen. This folding may be termed parasitic. The nature of the folding is related to its position on the overfold (Fig. 13.1(b)). This minor folding, frequently seen in the field, is of such a

(a)

(b)

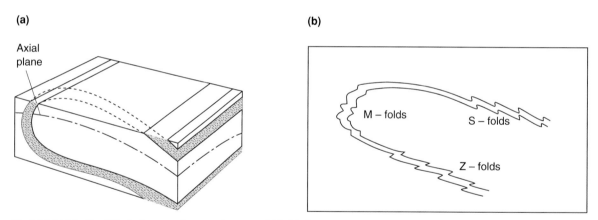

Fig. 13.1 (a) Idealised block diagram of a recumbent overfold. (b) The relationship of minor folds to their position on the overfold: M folds occur in the hinge region, S and Z folds are found on its limbs.

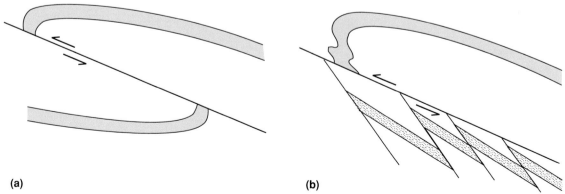

(a)   **(b)**

Fig. 13.2 (a) Section of a recumbent fold which has been thrust (a nappe) and (b) section showing imbricate structures which commonly occur beneath such a thrust plane.

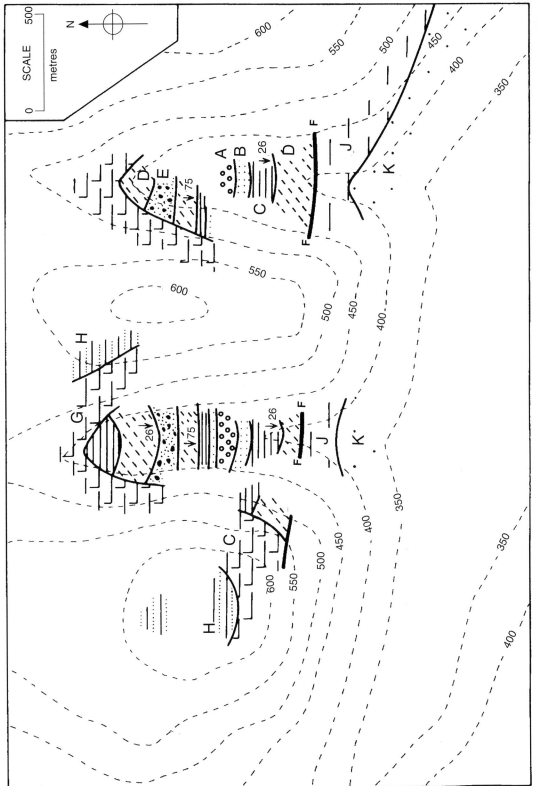

Map 44 Outcrops are few in the area portrayed by the map, but sufficient to enable structure contours to be drawn. Complete the outcrops over the whole map and draw a section along a north–south line. Briefly summarise the geological history.

scale that it would be revealed only on large-scale maps.

The axial plane of a recumbent overfold is nearly horizontal, but may be curved as in the above figure. Not infrequently the intense lateral stresses producing recumbent folds also cause rupture of the strata and produce a 'low-angle' reversed fault (making a low angle with the horizontal). The thrust recumbent fold is a nappe structure (Fig. 13.2(a)).

Thrust faults

A thrust is a low-angled fault plane along which movement has taken place, the strata above the thrust plane having been carried often for great distances in a near-horizontal direction, by intense earth movements, over the strata beneath. The thrust strata may have been displaced for a distance of many kilometres and may, for example as in the case of the Moinian rocks of Assynt (Sutherland) above the Moine Thrust, be quite different from any of the other rocks in the same district. On the other hand, the strata above the thrust may be similar to those beneath the thrust, but it should be noted that frequently older rocks might be thrust over younger ones. Thus it is possible to find Pre-Cambrian rocks overlying (due to thrusting) Cambrian rocks.

The strata above a thrust plane may be approximately parallel to it but, on the other hand, their dips may be unrelated to the inclination of the thrust plane, a whole block of rocks having been moved en masse. Frequently, just above a thrust plane, dips are locally affected by the movement along the thrust, beds being overturned due to the effects of drag over the rocks beneath the thrust.

The strata beneath a thrust plane may be very greatly affected by the forces associated with the thrusting. The effect on these beds is to produce imbricate structures. These comprise many parallel or near-parallel faults, sometimes of high angle of dip, although they are reversed faults that divide the unthrust area or foreland into 'slices' (Fig. 13.2(b)).

Axial plane cleavage

When rocks that are bedded (possessing what may be termed primary structures) are subjected to pressures – or stresses – they may, as we have discussed, become fractured by faults or they may become crumpled into fold structures. The stress applied to a rock may also cause it to be deformed and new structures are formed such as cleavage and schistosity. Cleavage is developed as a result of shortening of the rock in a direction perpendicular to the cleavage planes with a stretching or extension of the rock in the plane of the cleavage.

Cleavage is often found in rocks that are folded. The greatest shortening of strata due to folding is perpendicular to the fold axial planes and it therefore follows that cleavage planes are parallel to the fold axial planes; hence the cleavage is called axial plane cleavage. (This will not be true for a sequence of beds of different competency or, of course, in the case of complex refolded folds.)

Cleavage dips, indicated by a symbol such as ⬎, should not be confused with dips of bedding planes – to which there may not seem at first sight to be an obvious relationship. This information must not, of course, be dismissed as something clouding the issue: if the cleavage is axial plane cleavage (the most frequently found) there will be a consistent relationship between the cleavage direction and the main structural features (folds). The cleavage will be parallel to the axial planes of the folds.

The relationship between the cleavage dip and the dip of the bedding reveals, in overfolds, in which limb the beds are the right way up and in which limb the beds are inverted. Cleavage dip steeper than bedding = right way up; bedding dip steeper than cleavage dip = inverted (Fig. 13.3). This 'rule' is necessarily true when there has been only one period of folding (see Plate 31).

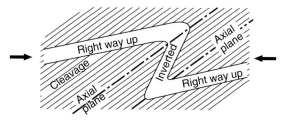

Fig. 13.3 Section showing cleavage/bedding relationships in overfolds with axial plane cleavage.

Plate 31 Axial plane cleavage and bedding, West Angle Bay, Pembrokeshire. These beds of Carboniferous shale, interbedded with thin limestones, dip steeply to the right (southwards). The near-vertical axial plane cleavage is prominent in the shales and indicates that these rocks form one limb of an upright fold (a syncline in this case).

Description of a geological map

All sources of information available should be used and coordinated: the information deducible from the map itself, the column of strata usually provided and sections showing the geological structures (the latter if not given on the map, drawn by the student). The description of a map should summarise the strata present. The general structural pattern and the trend of the chief structural features should be deduced from outcrop patterns. The broad relationship of topography to geology should be noted. It is usual to summarise the economic geology. Most important, a geological history of the area should be described. This comprises an attempt to date all geological events

relatively, the deposition of sediments, breaks in the succession (unconformities), the intrusion of igneous rocks and the development of faults and folds. The evolution of superficial deposits, the drainage pattern and the topography complete the history. In some cases a complete chronology cannot be deduced from map evidence and some events cannot be dated but their possible age and the ambiguity of the evidence must be discussed. Often, it is useful to illustrate your answer with a structural sketch map. An example is given of the main deductions that may be made from Map 45.

The geological history of Map 45

The earliest bed present is the ashy slate – formed by tuff falling into a sea in which argillaceous sediment was being deposited. A lull in vulcanicity but with continued deposition accounts for the overlying strata, now slate. The age of both formations is in doubt: they are thrust over Carboniferous Basal Conglomerate (which is clearly younger) and they are faulted against volcanic rocks. Though metamorphism itself is no guide to antiquity of strata, these slates (and ashy slates) may be the oldest strata present, since no other beds have been metamorphosed, and are thus older than the volcanics, which are earlier than the Ordovician grit. After the volcanic episode (perhaps prolonged although the thickness of volcanic rocks is not deducible since their structure is unknown) a conformable sequence of sediments was laid down. This comprises Ordovician grit and mudstone and Silurian sandstone. The lowest of these appears to rest partly on the microgranite, which must therefore have been intruded at an earlier date. The age of the east–west fault must be later than the slates and the volcanics that it cuts, but how much later cannot be established.

Dolerite dykes cut volcanic rocks and beds as young as the Silurian sandstone, and hence were intruded after the deposition of these beds. Although not cutting Carboniferous rocks the dykes may be contemporaneous with the post-Carboniferous sill, also of dolerite.

After (probably prolonged) non-deposition (since no Devonian age strata are seen). Carboniferous seas spread into the area depositing initially a basal conglomerate, though its base is not seen and we do not know what it rests on. As the sea became clearer and deepened, limestone was deposited, followed by the Yoredale beds – cyclic sediments laid down in shallow marine to terrestrial conditions. Two major post-Carboniferous events occurred: the thrusting northwards of older rocks over the Carboniferous, accompanied by overfolding of strata above the thrust plane, and the intrusion of the thick dolerite sill. Neither can be dated precisely. Both may be approximately the same age, referable to the late Carboniferous orogeny (the Variscan) although either could be of much later date. A northerly tilt was imparted to the area since the horizontally deposited Carboniferous strata have a northerly dip.

Mesozoic and Tertiary events are unknown since no strata of this age occur here. However, uplift and subsequent erosion have given rise to the present topography, southwards-sloping valleys cutting back into a high east–west escarpment. The topography relates closely to the underlying geology. Superficial deposits are not shown on this map.

Exercise on geological survey map

1. Assynt: 10 Geological Survey map (Special Sheet). How may the thrust planes be distinguished from the unconformities? Find on the map some examples of imbricate structures.

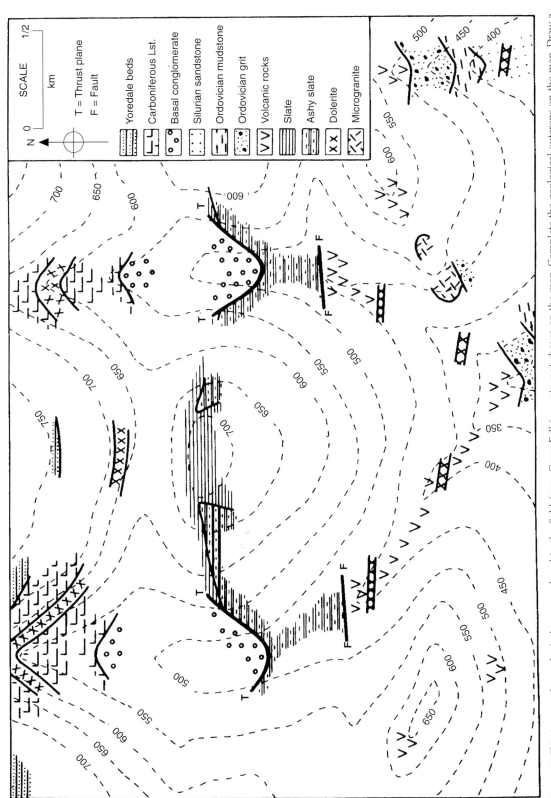

Map 45 This is part of a geological map produced in the field (near Cross Fell in northern England). Actual outcrops and exposure of geological boundaries are limited because of vegetation and the presence of superficial deposits of peat, alluvium, etc. not shown on the map. Complete the geological outcrops on the map. Draw a section along a north–south line to illustrate the structures. Write a geological history of the area including notes on the relative ages of the igneous rocks.

Appendix

a. True scale
1 cm = 500 m

b. ×5 Vertical
Exaggeration
1 cm = 100 m

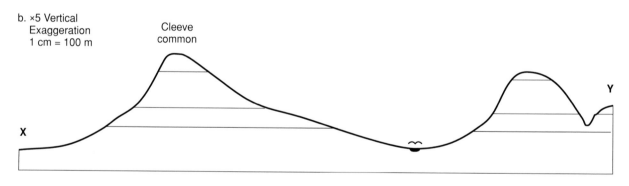

Figure A1 Geological sections of Map 3.

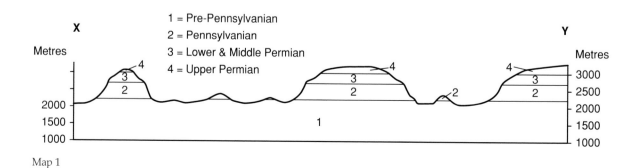

Map 1

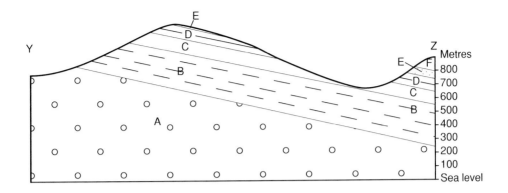

Map 5

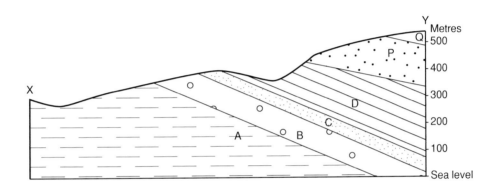

Map 9

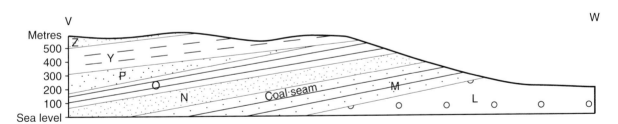

Map 10

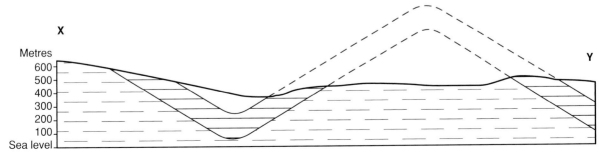

Map 11

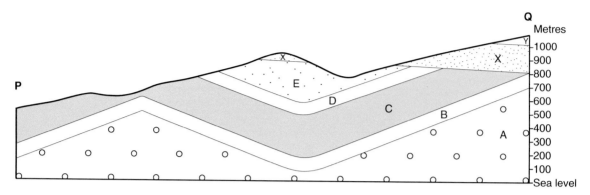

Map 12

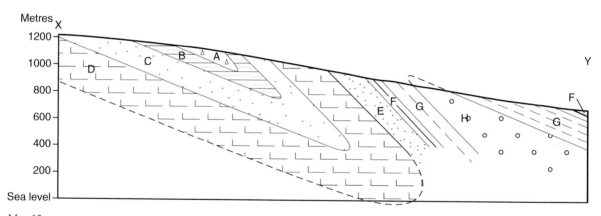

Map 13

Map 14

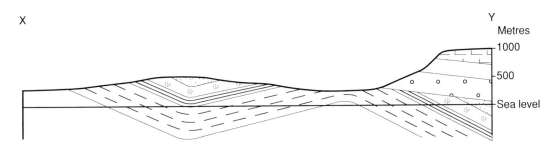

Section profile X–Y, Map 15.

Map 16

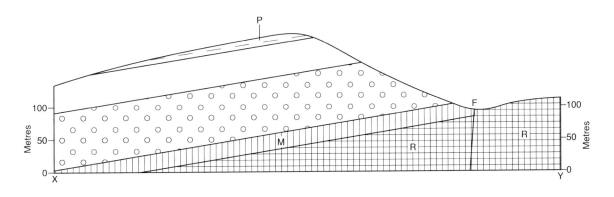

Map 18

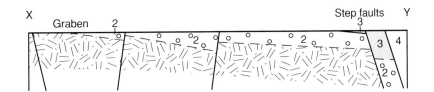

Map 19

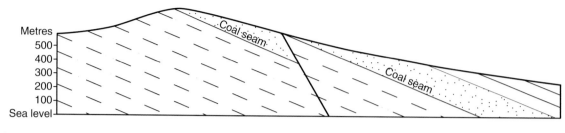

Map 20

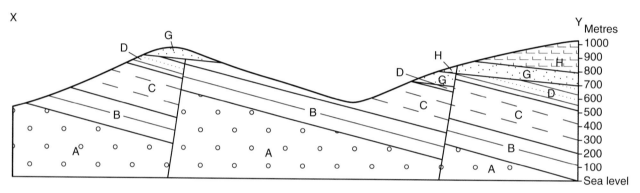

Map 21

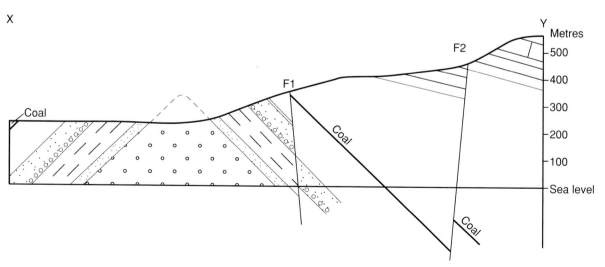

Map 22

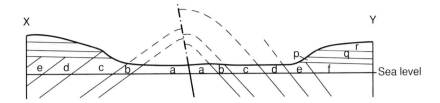

Map 23

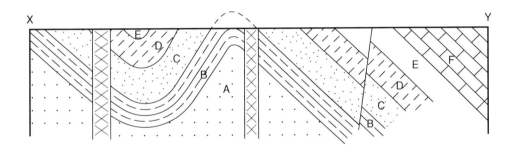

Map 24

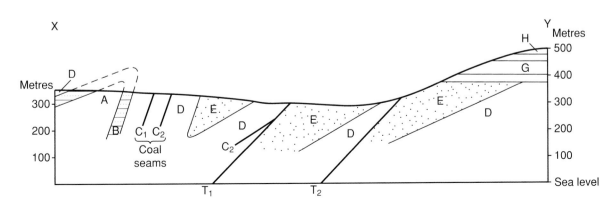

Map 25

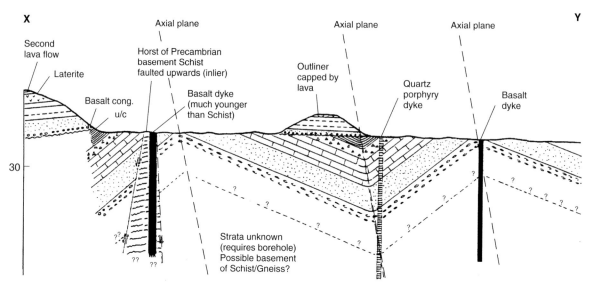

The sequence consists of basement crystalline rocks, exposed in faulted inliers, overlain by a thick sequence of sediments folded into asymmetric synclines and anticlines (axial planes at 80°) with NE–SW axes (probably Caledonian in origin). Shallow dipping or horizontal deltaic sediments and lava flows lie unconformably on the folded sequence beneath. Dykes of igneous rock are intruded into the sequence last.

POSSIBLE SOLUTION TO SECTION X–Y, MAP 26

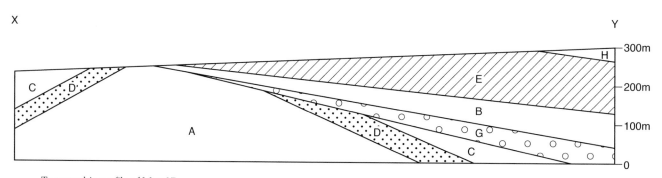

Topographic profile of Map 27.

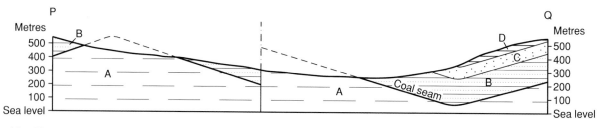

Map 28

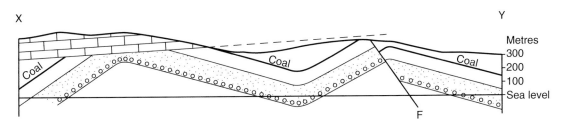

Map 29

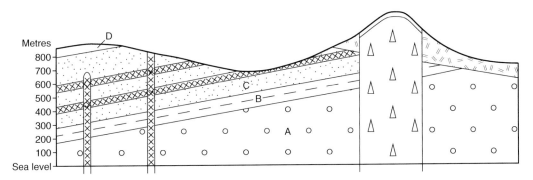

Map 31

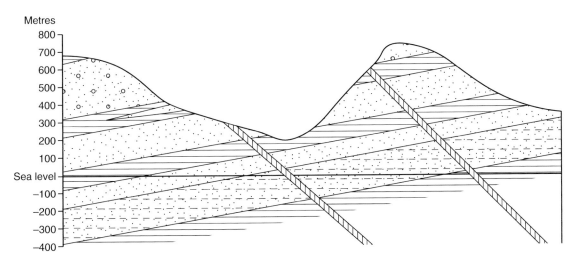

Map 32

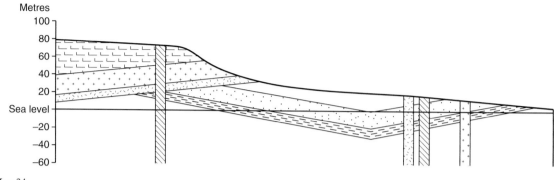

Map 34

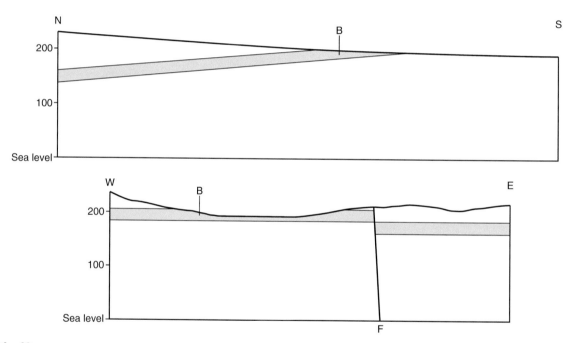

Map 39

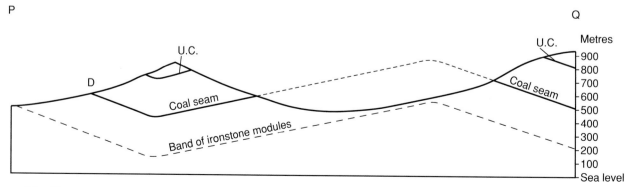

Map 42

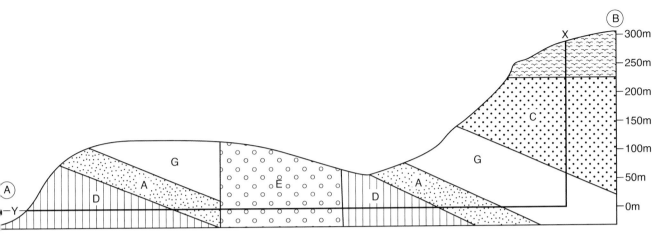

Topographic profile of Map 43.

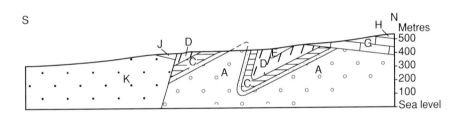

Map 44

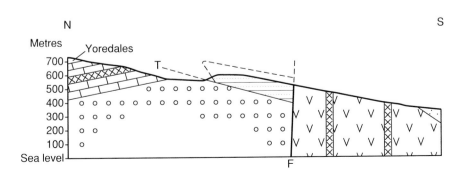

Map 45

COMPLETED MAP SOLUTIONS

Three-point problems and economic problems

(Maps reduced to approximately 70%, except Map 42 full size)

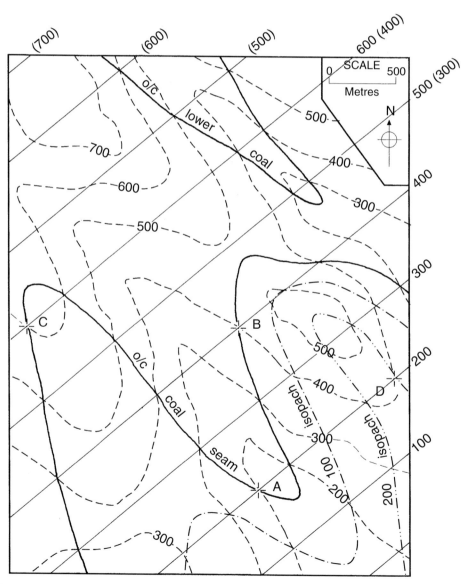

Map 6 Solution

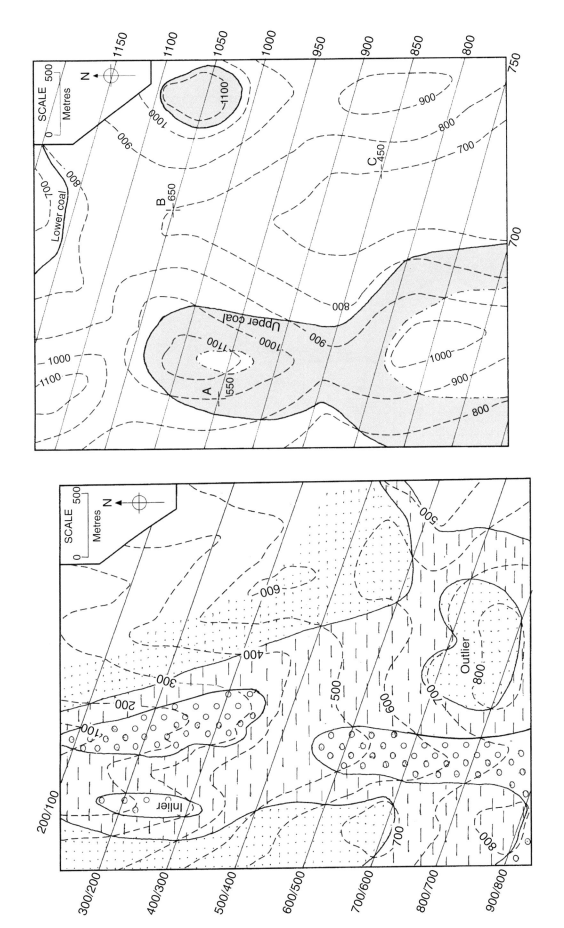

Map 7 Solution

Map 8 Solution

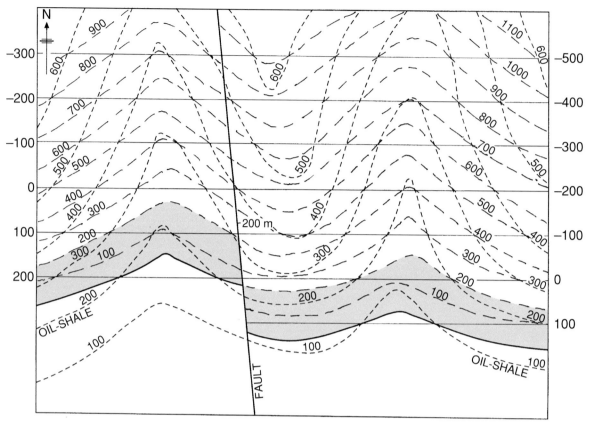

Map 38 Solution

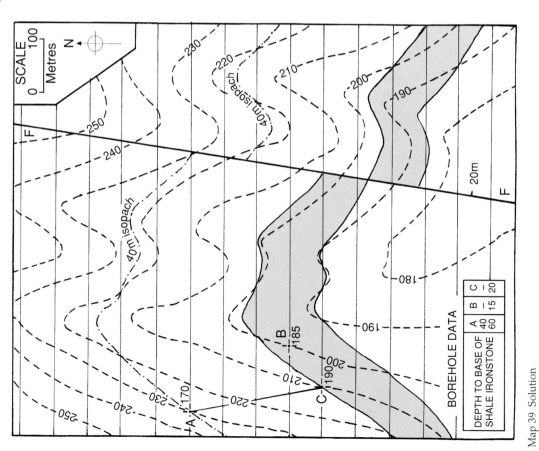

Map 39 Solution

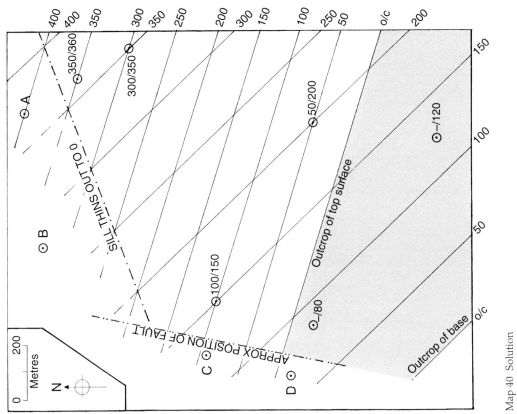

Map 40 Solution

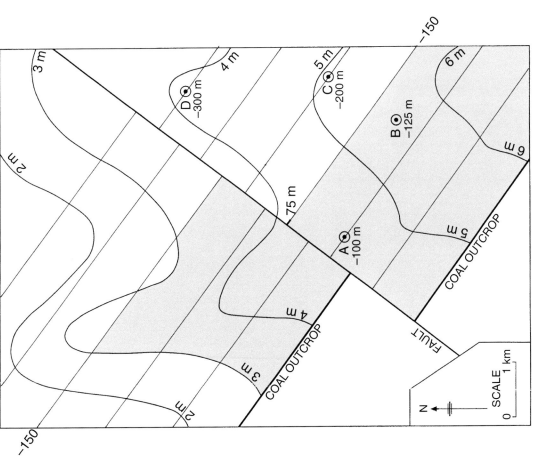

Map 41 Solution

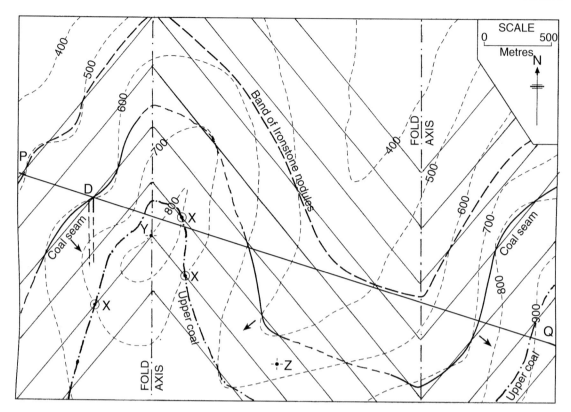

Map 42 Solution

Numerical Answers

Map 2 Sandstone 2 — 100 m
Mudstone — 150 m
Shale — 50 m
Sandstone 1 — 150 m

Map 4 1 in 2½; Bearing 174°
(6°E. of South)

Map 5 1 in 5; East
B — 250 m
C — 100 m
D — 100 m
E — 50 m

Map 6 200 m

Map 10 A — 450 m
B — 200 m
C — absent

Map 17 500 m
200 m

Map 28 Wrench fault
No throw
Lateral displacement
3.2 cm = 615 m

Map 42 Y — 400 m
Z — 110 m
1 in 4 (14°)

Answers to Map Questions

Map 30
1. Dykes dip steeply east.
2. Pillow lavas, with shallow dips to the WSW, outcrop to the west of the dykes.
3. Gabbroic intrusions outcrop to the east of the dykes dipping steeply to the east or ENE.
4. The mélange
5. Random fragments of the sheeted dolerite dykes
6. Contact X is an unconformity
   Contact Y is an intrusive contact
7. (a) Western outcrop of pillow lavas
   (b) Towards the east
   (c) Middle section of the map within the E–W trending fault zone
   (d) Approximately E–W parallel to the transform fault system

Map 33
1. (a) Between the conglomerate and the older sequence of rocks.
   (b) Rock D
   (c) Rock B
   (d) The limestone-conglomerate succession
   (e) The faulted NW–SE trending dyke in the centre of the map area

(f) A faulted plunging syncline with an axial trace trending ENE–WSW across the centre of the map with the shale horizon at its core
2. Downthrow of the major fault is on its eastern side
3. Bed B dips at a steeper angle in the north. Bed B dips due south on this limb of the fold
4. This area displays a ring complex with Granite I as the oldest intrusion, a thin older ring dyke of dolerite and a younger ring dyke formed of Granite II outcropping in the centre (See Fig.11.4 Erosion Level 2)

Map 36
1. (a) Pre-existing rocks strike NE–SW
   (b) Point of Impact is at the centre of the double-rimmed crater
   (c) Rock A is the ejecta blanket
   (d) Impacting body arrived from a north-westerly direction
2. The freshwater limestone (Rock C) was laid down in the shallow lake which occupied the resultant impact crater and is constrained by the outer crater rim

3. Age of Impact is between late Cretaceous and mid-Tertiary times

Problem 1

3. The axial trace of the anticline is aligned along a bearing of 275° whereas the syncline to the north trends at 290°. The fault on the southern limb of the syncline downthrows to the north.

4. The components of the trellised drainage diagram (Figure 6:8) are as follows:

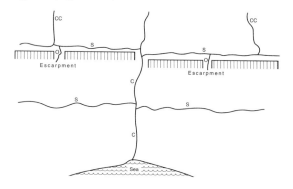

Problem 3

1. X is a plutonic intrusion
   Y is a dyke
   Z is a lava flow

2. Using only the evidence available on the section:
   A Ordovician
   C Devonian
   D Between Carboniferous and Tertiary (Eocene) in age

3. Pillow lava B is "right way up" as its associated contact metamorphism has affected Shale A below

4. Rock Z is likely to be a flow-folded rhyolite lava

5. Rock E are xenoliths formed by magmatic stoping of enclosing rock A

6. Boundary N–N is an angular unconformity which shows the planation of a former landscape (including an igneous intrusion) and also the resultant included fragments of the igneous rock along the weathered surface.

Table 1 True and apparent dip

The apparent dip of strata, when viewed at an angle to the dip direction, is always less than the true dip. The table below sets out apparent dips at regular intervals. Intermediate values can be interpolated with the graph provided. An example is shown in the table: a true dip of 15° (the left hand column) viewed at 60° to the true dip direction (the top row) appears as a dip of 7.6° (in box).

Dip	Angle of horizontal deviation from true dip direction													
	0	10	20	30	40	50	55	60	65	70	75	80	85	90
0	0	0	0	0	0	0	0	0	0	0	0	0	0	0
5.0	5.0	4.9	4.7	4.3	3.8	3.2	2.9	2.5	2.1	1.7	1.3	0.9	0.4	0.0
10.0	10.0	9.9	9.4	8.7	7.7	6.5	5.8	5.0	4.3	3.5	2.6	1.8	0.9	0.0
15.0	15.0	14.8	14.1	13.1	11.6	9.8	8.7	7.6	6.5	5.2	4.0	2.7	1.3	0.0
20.0	20.0	19.7	18.9	17.5	15.6	13.2	11.8	10.3	8.7	7.1	5.4	3.6	1.8	0.0
25.0	25.0	24.7	23.7	22.0	19.7	16.7	15.0	13.1	11.1	9.1	6.9	4.6	2.3	0.0
30.0	30.0	29.6	28.5	26.6	23.9	20.4	18.3	16.1	13.7	11.2	8.5	5.7	2.9	0.0
35.0	35.0	34.6	33.3	31.2	28.2	24.2	21.9	19.3	16.5	13.5	10.3	6.9	3.5	0.0
40.0	40.0	39.6	38.3	36.0	32.7	28.3	25.7	22.8	19.5	16.0	12.3	8.3	4.2	0.0
45.0	45.0	44.6	43.2	40.9	37.5	32.7	29.8	26.6	22.9	18.9	14.5	9.9	5.0	0.0
50.0	50.0	49.6	48.2	45.9	42.4	37.5	34.4	30.8	26.7	22.2	17.1	11.7	5.9	0.0
55.0	55.0	54.6	53.3	51.0	47.6	42.6	39.3	35.5	31.1	26.0	20.3	13.9	7.1	0.0
60.0	60.0	59.6	58.4	56.3	53.0	48.1	44.8	40.9	36.2	30.6	24.1	16.7	8.6	0.0
65.0	65.0	64.7	63.6	61.7	58.7	54.0	50.9	47.0	42.2	36.3	29.0	20.4	10.6	0.0
70.0	70.0	69.7	68.8	67.2	64.6	60.5	57.6	53.9	49.3	43.2	35.4	25.5	13.5	0.0
75.0	75.0	74.8	74.1	72.8	70.7	67.4	65.0	61.8	57.6	51.9	44.0	32.9	18.0	0.0
80.0	80.0	79.8	79.4	78.5	77.0	74.7	72.9	70.6	67.4	62.7	55.7	44.6	26.3	0.0
85.0	85.0	84.9	84.7	84.2	83.5	82.2	81.3	80.1	78.3	75.7	71.3	63.3	44.9	0.0
90.0	90.0	90.0	90.0	90.0	90.0	90.0	90.0	90.0	90.0	90.0	90.0	90.0	90.0	90.0

Table 2 Bed thickness, if vertical thickness of bed (V.T.) = 100 m

Angle of dip (degrees)	True thickness of bed (m)
0	100
10	98.5
20	94.0
30	86.6
40	76.6
50	64.3
60	50.0
70	34.2
80	17.4

Table 3 Outcrop width if true thickness of bed = 100 m

Angle of dip (degrees)	-Outcrop width on a horizontal surface
0	infinite
10	575.9
20	292.4
30	200.0
40	155.6
50	130.5
60	115.5
70	106.4
80	101.5
90	100.0

Each of the following eleven diagrams may be cut out and folded to form a box. You then have a three-dimensional model of a geological structure, each side of the block showing a section through the structures outcropping on the surface.

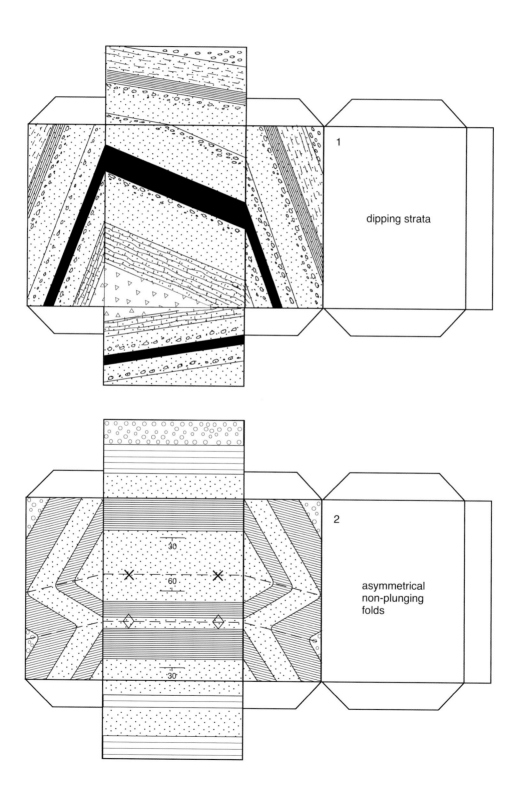

1

dipping strata

2

asymmetrical
non-plunging
folds

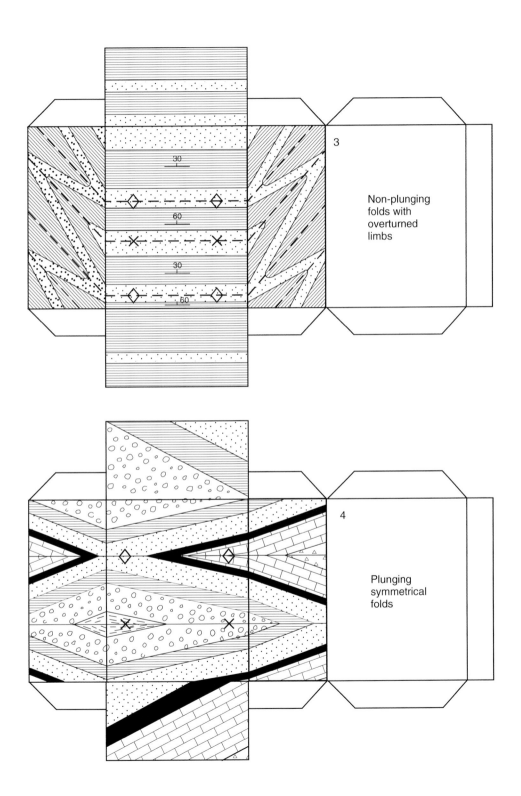

3

Non-plunging
folds with
overturned
limbs

4

Plunging
symmetrical
folds

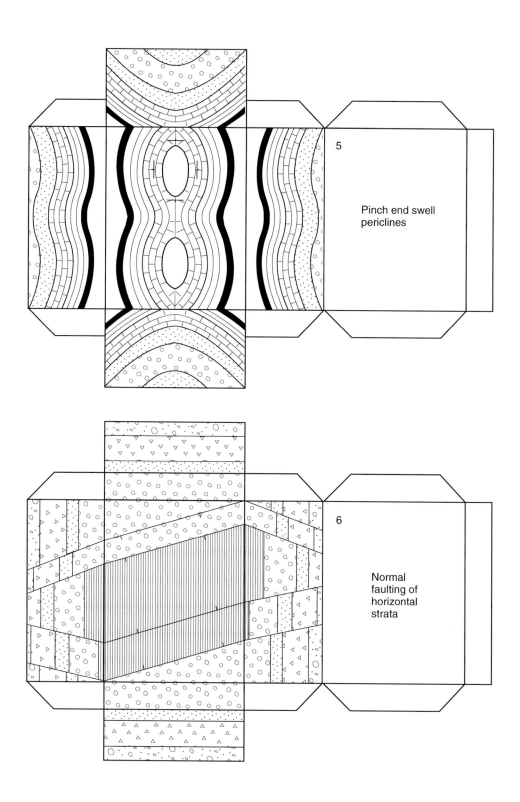

5

Pinch end swell periclines

6

Normal faulting of horizontal strata

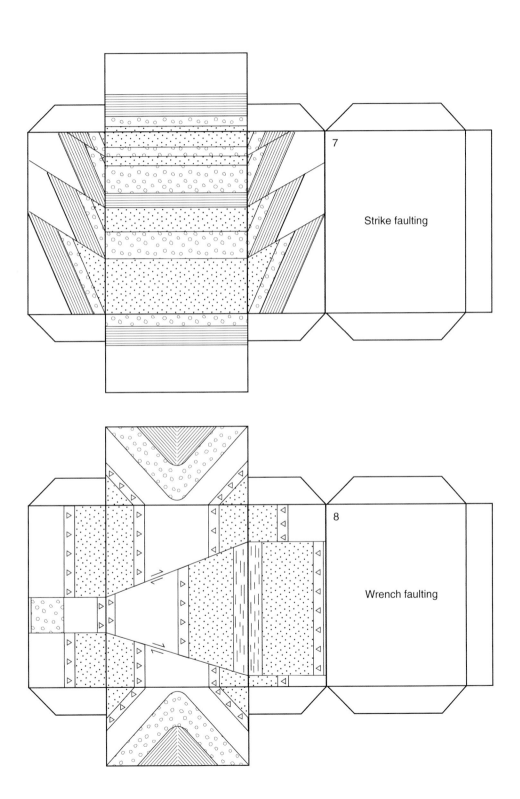

7 Strike faulting

8 Wrench faulting

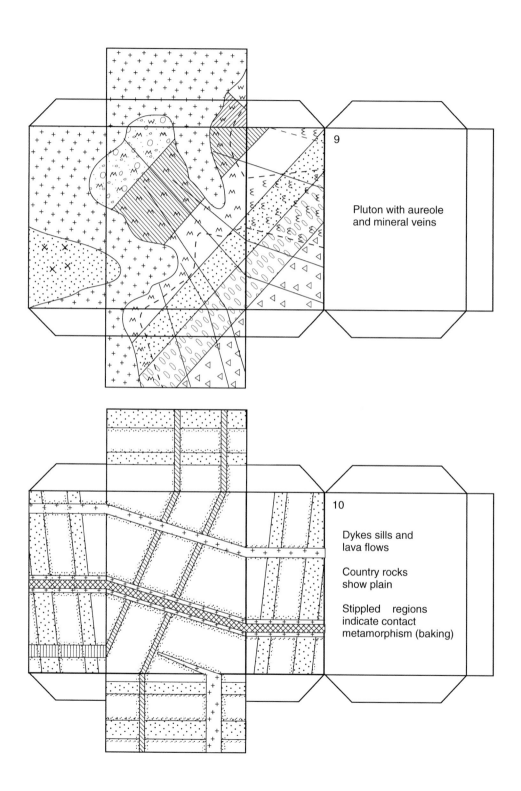

9

Pluton with aureole
and mineral veins

10

Dykes sills and
lava flows

Country rocks
show plain

Stippled   regions
indicate contact
metamorphism (baking)

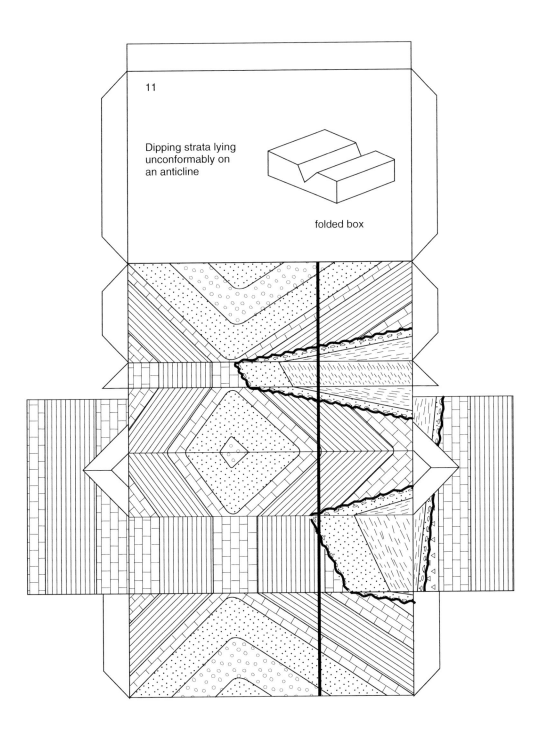

11

Dipping strata lying
unconformably on
an anticline

folded box

Glossary

absolute dating Method using the decay of radionuclides within certain rock types (mainly igneous) which allow a numerical age in millions of years to be determined.

acid Describes igneous rocks containing more than 66 per cent silica.

agglomerate A pyroclastic volcanic rock formed of blocks (derived from older lavas from within the volcano or rocks from beneath the volcano) together with fragments of more than 50 mm diameter that have been ejected in a plastic state.

angular unconformity Represents a period of erosion of a series of sedimentary layers which has been folding and faulted prior to being eroded and covered by younger layers. The beds beneath such an unconformity are therefore not parallel with those above.

asteroid Small rocky or metallic objects within the Solar System most of which lie within the asteroid belt between the orbits of Mars and Jupiter.

astrobleme An eroded impact crater on the Earth's surface recognisable by its geological structure and highly shocked rock types and impact melts.

autobrecciation A breccia layer formed of lava clasts produced at the surface of acid or intermediate lavas and caused by the break up of the cooling crust by the movement of the magma beneath.

axial plane A planar surface which connects the hinge lines of the strata involved within a particular fold structure.

axial trace The intersection of the axial plane of a fold with the earth's surface.

basalt A fine-grained basic lava or minor intrusion composed mainly of calcium-rich plagioclase feldspar and pyroxene with or without olivine.

basic Describes igneous rocks more than 52 per cent silica.

batholith A large, composite, intrusive, igneous rock body formed of numerous plutons which can be several hundred kilometres long and a hundred kilometres wide.

brittle deformation Describes a response to regional stress whereby rock materials undergo little or no plastic deformation before rupture (faulting) occurs.

columnar jointing Formed by contraction during the slow cooling of lava flows and minor intrusions consisting of sets of fractures forming parallel prismatic columns within the rock mass.

comet A small body, composed of ice and dust, in orbit around the Sun, which represents a relict from the formation of the Solar System.

competent Used to describe sedimentary rocks which are able to withstand the stresses of folding without showing any flowage or changes in their original thickness.

country rock The rock immediately in contact with an intrusive mass of igneous rock.

cross-bedding An internal arrangement within a stratified sediment characterised by parallel sloping minor layers deposited at an angle to the main stratification.

cuesta An asymmetric ridge formed by dipping layers of rock with a steep cliff on one side cutting across the layers and a gentle slope on the other side parallel to the stratification.

cupola A mass of plutonic igneous rock which extends upwards towards the surface from the overall level of a major batholith.

dacite A dark-coloured volcanic rock, often showing flow banding, which represents the extrusive equivalent of granodiorite.

detrital Describes mineral fragments that have been transported and deposited as sediments from the erosion of pre-existing rocks.

dolerite Medium grained igneous rock forming the hypabyssal equivalent of basalt formed mainly in dykes and sills.

drag fold A fold that develops in rock strata adjacent to a fault during or just before movement.

ductile deformation Describes a response to regional stress whereby rock materials undergo considerable plastic deformation before rupture (faulting) occurs.

ejecta blanket Material thrown out by crater-forming impact which covers the area immediately outside the crater rims.

felsite A light coloured fine grained igneous rock composed mainly of quartz and feldspar.

flame structure Flame-shaped intrusion of argillaceous sediment which has been squeezed upwards into an overlying layer of coarser sediment.

flow banding Alternating layers of different texture and /or composition formed as a result of flowage within a magma.

gabbro A coarse-grained intrusive, basic igneous rock.

Galilean satellites Four largest moons of Jupiter (Io, Europa, Ganymede and Callisto) first discovered by Italian astronomer Galileo Galilei in 1610.

graded bedding Displayed by clastic sediments whereby the maximum grain size progressively decreases from the base to the top of a particular bed.

granite A coarse-grained, intrusive acid igneous rock.

granodiorite Coarse-grained, plutonic igneous rock containing less alkali-feldspar than the average granite.

half-life The time taken for half a group of a certain isotope to decay.

hardground Represents a hiatus in sedimentary deposition within a particular rock sequence. Commonly shows the activities of burrowing worms or bivalves as well as chemical changes such as phosphatization.

hinge The portion of a fold where the curvature is at its greatest.

hypabyssal Refers to any igneous rock of medium grain size which crystallised at relatively shallow depths often in the form of minor intrusions.

ignimbrite Rock formed by the solidification of deposits from pyroclastic flows.

impactite A composite rock type of meteoritic material fused with melted rock formed by crater formation on a planetary surface.

incompetent Used to describe sedimentary rocks which, within a folding event, demonstrate flowage together with thickening and thinning around the fold profile.

intermediate Describes igneous rocks that are compositionally transitional between acid and basic (52 – 66 per cent silica).

intrusive contact The boundary between the enclosing country rock and an intrusive igneous rock.

island arc A volcanic island chain that forms where an oceanic plate subducts beneath another continental or oceanic plate.

isotope Atoms of an element which have the same atomic number but a different atomic weight.

lacustrine Related to or produced by lakes.

load cast Formed at the interface between two sedimentary layers in which the coarser uppermost sediment protrudes down into the finer sediment below.

Ma Million years ago (abbreviation).

magma Generic term to describe molten rock usually of silicate composition but occasionally consists of carbonate or elemental sulphur.

metasomatism The alteration of existing minerals by their interaction with capillary solutions emanating from outside or within the rock mass leading to the crystallisation of new minerals.

meteorite A piece of rock or metal alloy from space that survives passage through the atmosphere to strike the Earth's surface.

obsidian An igneous rock consisting of a solid mass of volcanic glass, usually black, but sometimes brown or red and usually formed from an acid rhyolitic magma.

ophiolite A segment of oceanic crust consisting of a suite of basic and ultrabasic igneous rocks, associated with pelagic sediments, emplaced within the continental crust through plate collision events.

palaeosol An ancient soil preserved within the stratigraphic record.

palimpsest An impact crater on an icy surface whose initial sharp crater rims have been smoothed by glacier-like surface flow.

parallel unconformity Represents a period of erosion or prolonged non-deposition within a sedimentary sequence which separates two beds showing parallel bedding.

pelagic Referring to sediments which accumulate on the floor of the open sea or ocean, well beyond the direct influence of the land, and comprise materials which originate almost exclusively from within the deep water environment.

piecemeal stoping A process occurring at the margins of both major and minor intrusions whereby the magma displaces fragments of country rock which then sink or are assimilated into the hot magma.

pillow lava Characterised by a mass of rounded and slightly flattened lava bodies resembling pillows. Formed usually in basalt lavas extruded into a subaqueous environment.

plunge The inclination of a fold axis, measured in the vertical plane.

pumice A frothy, glassy lava, often fragmented, which is light enough to float on water. Common in pyroclastic fall deposits and within pyroclastic flows.

pyroclastic fall Pyroclastic material exploded from a volcanic vent which has passed through the air before being deposited onto nearby land or into the sea.

pyroclastic flow A hot, particulate, mass of pyroclastic material moving as a surface flow away from the volcanic vent. Flow is controlled by gravity but may be partially fluidised.

radioactive decay The process by which a radioactive atomic species undergoes fission or releases particles.

radionuclide An unstable nucleus of an atom that undergoes spontaneous radioactive decay thereby emitting radiation and eventually changing from one element into another.

relative dating Methods using fossil assemblages and sedimentary features which allow the age of one geological horizon to be determined relative to another.

rhyolite An acid, extrusive igneous rock which commonly shows flow banding.

roof pendant Formed of downward projections of country rock into the surface of a major batholith. Found between major cupolas on a batholith's upper surface.

spilite An altered basalt, usually pillow lava, in which new minerals have formed caused by the metasomatic interchange with heated sea water.

stratosphere The layer of the atmosphere which extends on average between 10 km and 50 km above the Earth's surface.

tension joints A surface fracture within competent rocks which shows no displacement and which forms around the hinge line of both anticlines and synclines.

trap topography A landscape formed by a sequence of horizontal or near horizontal lava flows separated by more weathered, thinner layers of palaeosols formed by subaerial weathering of the lava surfaces.

tuff The consolidated product of volcanic ash fall or ash flow (see ignimbrite) forming a fine grained pyroclastic rock.

turbidity current A dense, subaqueous flow of water containing suspended sediments which moves downslope by gravity.

viscosity Property shown by a substance that offers internal resistance to the force tending to cause the substance to flow.

way-up criteria Features that allow the original altitude of a rock mass to be determined and thus the direction of younging.

welding The process by which hot, plastic pyroclastic fragments are agglutinated by retained heat and load pressure within the base of a pyroclastic flow.

xenolith Fragments of older rocks carried along within lava flows or relics of country rock found in the margins of intrusive igneous rocks.

younging The direction in which a succession of rocks becomes younger.

Index

Page numbers in **bold** refer to figures and plates.